THE DOLLAR H

THE CLASSIC HANDBOOK OF AMERICAN FREE-RANGE EGG FARMING

MILO M. HASTINGS

FORMERLY POULTRYMAN AT
KANSAS EXPERIMENT STATION;
LATER IN CHARGE OF THE COMMERCIAL
POULTRY INVESTIGATION OF
THE UNITED STATES
DEPARTMENT OF AGRICULTURE

A Norton Creek Classic
EDITED BY ROBERT PLAMONDON

NORTON CREEK PRESS
36475 NORTON CREEK RD.
BLODGETT OR 97326
http://www.plamondon.com
nortoncreek@plamondon.com

ORIGINALLY PUBLISHED IN 1909
BY THE ARCADIA PRESS

The Dollar Hen
The Classic Handbook of American Free-Range Egg Farming.

ISBN 0-9721770-1-9

2 4 6 8 10 9 7 5 3 1

Introduction to the
Norton Creek Edition

Milo Hastings' *The Dollar Hen* changed my life. The book was almost 90 years old when I first read it, but it set me on the path to a successful free-range egg business. It gave me more good advice than any poultry book I've read, before or since. But Hastings was always ahead of his time.

Hastings was Poultryman at the Kansas Experiment Station in 1902, only a few years after any scientist anywhere turned his attention to practical poultry questions, but he was given no funds. He moved on to the USDA, which charged him with learning all about the commercial poultry industry as it then stood, which brought some much-needed practical information into the field.

In 1919, he wrote a classic work of science fiction, *City of Endless Night*. Because of Hastings' practical bent, the book is chillingly plausible, which isn't something you can say of most SF works of that era.

In the Twenties, Hastings was active in the physical culture movement, writing books and serving as food editor of *Physical Culture* magazine. The movement was mostly focused on the health benefits of exercise, but Hastings added a much-needed emphasis on nutrition. This was when vitamins were a new concept. Hastings was right there with both theory and practice. He wrote books on nutrition and an early work on high blood pressure.

Hastings was that rarest of creatures, the practical philosopher. He wanted to make sense of the world in a way that would be immediately useful to his readers.

When I started raising chickens on my Oregon farm in 1996, I wanted to learn about profitable free-range egg farming. I have always tried to turn my hobbies into businesses, and I saw no reason why farming should be an exception.

The modern literature I could find on poultry farming fell into three categories:

1. Professional literature for modern factory farms. This information is interesting and often useful, but is generally can't be applied directly to the problems of small farms.
2. Literature for backyarders, fanciers, and hobbyists. Though fascinating and in many cases charming, these works are generally quite useless for practical farmers. They are written by and for people with very small flocks who look forward to spending time and money on their fascinating hobby without any real expectation of profit. To the farmer, profit is what pays the mortgage, puts the kids through college, and pays for retirement (or, in my case, make a part-time contribution to these things), and any work that treats profit as an optional extra provides hazardous guidance.
3. Literature motivated by politics, a romantic view of farm life, or a fear of chemicals. I found these particularly misleading, because they tended to represent theories as facts and wishful thinking as established practices. It turns out that most of this writing is done by non-farmers, though they do not advertise this fact.

To learn something practical, I turned to the poultry literature of yesteryear, when small farms like mine were the norm, and a flock of hens figured into nearly every farm in the country. I quickly learned that the period of interest started around 1900, when practical poultry research began, and ended around 1960, when the switch to factory farming was complete.

I live close to Oregon State University, where just about every poultry book ever written can be found on the first floor of the Valley Library. However, the first dozen or so books I read were just as impractical as their modern equivalents. There have always been books written about the delights of country life by newly transplanted city folks. Ominously, many were written by people whose poultry business was in its second or third year of operation, before the owner's money or luck had been given time to run out.

In short, my search for examples of practical, unpretentious, profitable free-range egg farming was coming up empty.

Then I found *The Dollar Hen*.

Hastings was refreshingly practical, even cynical. I liked him immediately. He was the only poultry author I had found who came right out and said that a farmer's time is too valuable to waste, and that results come not from fancy methods but from simple ones, intelligently applied.

More than that, he had investigated every aspect of the poultry industry when working for the U.S. Department of Agriculture, and he explained the fundamentals of how it all worked. Because fundamentals never change, this information is as useful now as it was when it was written, almost a century ago.

The book is particularly useful to today's small farmers and those who deal with them, because the problems and methods of small farmers have changed relatively little, and have more in common with those of Hastings' day than they do with today's industrialized farms.

I have edited Hastings' original work, replacing terms that have fallen out of use with modern equivalents, adding explanatory footnotes, and knocking some of the rough edges off the original edition's erratic punctuation. At no point have I changed the meaning of a single sentence.

Robert Plamondon
Blodgett, Oregon
March, 2003

Why This Book Was Written

Twenty-five years ago there were in print hundreds of complete treatises on human diseases and the practice of medicine. Notwithstanding the size of the book-shelves or the high standing of the authorities, one might have read the entire medical library of that day and still have remained in ignorance of the fact that outdoor life is a better cure for tuberculosis than the contents of a drug store. The medical professor of 1885 may have gone prematurely to his grave because of ignorance of facts which are today the property of every intelligent man.

There are today, on the book-shelves of agricultural colleges and public libraries, scores of complete works on "Poultry" and hundreds of minor writings on various phases of the industry. Let the would-be poultryman master this entire collection of literature, and he is still in ignorance of facts and principles, a knowledge of which in better-developed industries would be considered prime necessities for carrying on the business.

As a concrete illustration of the above statement, I want to point to a young man, intelligent, enterprising, industrious, and a graduate of the best known agricultural college poultry course in the country. This lad invested some $18,000 of his own and his friends' money in a poultry plant. The plant was built and the business conducted in accordance with the plans and principles of the recognized poultry authorities. Today the young man is bravely facing the proposition of working on a salary in another business to pay back the debts of honor resulting from his attempt to apply, in practice, the teaching of our agricultural colleges and our poultry bookshelves.

The experience just related did not prove disastrous from some single item of ignorance or oversight; the difficulty was that the cost of growing and marketing the product amounted to more than the receipts from its sale. This poultry farm, like the surgeon's operation, "was successful, but the patient died."

The writer's belief in the reality of the situation as above portrayed warrants him in publishing the present volume. Whether his criticism of poultry literature is founded on fact or fancy may, five years after the copyright date of this book, be told by any unbiased observer.

I have written this book to assist in placing the poultry business on a sound scientific and economic basis. The book does not pretend to be a complete encyclopedia of information concerning poultry, but treats only of those phases of poultry production and marketing upon which the financial success of the business depends.

The reader who is looking for information concerning fancy breeds, poultry shows, patent processes, patent foods, or patent methods will be disappointed, for the object of this book is to help the poultryman to make money, not to spend it.

How To Read This Book

Unless the reader has picked up this volume out of idle curiosity, he will be one of the following individuals:

1. A farmer or would-be farmer who is interested in poultry production as a portion of the work of general farming.
2. A poultryman or would-be poultryman who wishes to make a business of producing poultry or eggs for sale as a food product or as breeding stock.
3. A person interested in poultry as a diversion and who enjoys losing a dollar on his chickens almost as well as earning one.
4. A man interested in poultry in the capacity of an editor, teacher or someone engaged as a manufacturer or dealer in merchandise the sale of which is dependent upon the welfare of the poultry industry.

To the reader of the fourth class, I have no suggestions to make save such as he will find in the suggestions made to others.

To the reader of the third class, I wish to say that if you are a shoe salesman who has spent your evenings in a Brooklyn flat, drawing up plans for a poultry plant, I have only to apologize for any interference that this book may cause with your highly fascinating amusement.

To the poultryman already in the business, or to the man who is planning to engage in the business for reasons equivalent to those which would justify his entering other occupations of the semi-technical class, such as dairying, fruit growing or the manufacture of washing machines, I wish to say it is for you that "The Dollar Hen" is primarily written.

This book does not assume you to be a graduate of a technical school, but it does bring up discussions and use methods of illustration that may be unfamiliar to many readers. That such matter is introduced is because the subject requires it; and if it is confusing to the student he will do better to master it than to dodge it. Especially would I call your attention to the diagrams used in illustrating various statistics. Such diagrams are technically called "curves." They may at first seem mere crooked lines, if so I suggest that you get a series of figures in which you are interested, such as the daily egg yields of your own flock or your monthly food bills, and plot a few curves of your own. After you catch on you will be surprised at the greater ease with which the true meaning of a series of figures can be recognized when this graphic method is used.

I wish to call the farmer's attention to the fact that poultry-keeping as an adjunct to general farming, especially to general farming in the Mississippi Valley, is quite a different proposition from poultry production as an independent business. Poultry-keeping as a part of farm life and farm enterprise is a thing well worth while in any section of the United States, whereas poultry-keeping, a separate occupation, requires special location and special conditions to make it profitable. I would suggest the farmer first read Chapter XVI, which is devoted to his special conditions. Later he may read the remainder of the book, but should again consult the part on farm poultry production before attempting to apply the more complicated methods to his own needs.

Chapter XVI, while written primarily for the farmer, is, because of the simplicity of its directions, the best general guide for the beginner in poultrykeeping wherever he may be.

To the reader in general, I want to say that the table of contents, a part of the book which most people never read, is in this volume so placed and so arranged that it cannot well be avoided. Read it before you begin the rest of the book, and use it then and thereafter in guiding you toward the facts that you at the time particularly want to know. Many people, in starting to read a book, find something in the first chapter which does not interest them and cast aside the work, often missing just the information they are seeking. The conspicuous arrangement of the contents is for the purpose of preventing such an occurrence in this case.

Contents

Chapter 1. Is There Money In The Poultry Business?.....14
A Big Business; Growing Bigger 14
Less Ham and More Eggs 17
Who Gets the Hen Money? 18

Chapter 2. What Branch Of The Poultry Business?. 21
Various Poultry Products. 21
The Duck Business 24
Squab Business Overdone. 24
Turkeys Not a Commercial Success 25
Guinea Growing a New Venture. 25
Geese—the Fame of Watertown 26
The Ill-omened Broiler Business. 26
South Shore Roaster 31
Too Much Competition in Fancy Poultry 32
Egg Farming the Most Certain and Profitable. 34

Chapter 3. The Poultry Producing Community. 36
Established Poultry Communities 38
Developing Poultry Communities. 40
Will Cooperation Work? 44
Cooperative Egg Marketing In Denmark 47
Corporation or Cooperation?. 48

Chapter 4. Where To Locate 50
Some Poultry Geography 50
Chicken Climate 56
Suitable Soil 58
Marketing—Transportation. 59
Availability of Water. 62
A Few Statistics. 62

Chapter 5. The Dollar Hen Farm 66
The Plan of Housing 66
The Feeding System 68
Water Systems. 69
Outdoor Accommodations 70
Equipment for Chick Rearing 71
Twenty-five Acre Poultry Farms. 76
Five Acre Poultry Farm. 84

Chapter 6. Incubation 88
Fertility of Eggs. 88
The Wisdom of the Egyptians. 92
Principles of incubation. 94
Moisture and Evaporation 98
Ventilation—Carbon Dioxide 105
Turning Eggs. 108
Cooling Eggs. 109
Searching for the "Open Sesame" of Incubation. 110
The Box Type of Incubator In Actual Use 112
The Future Method of Incubation 116

Chapter 7. Feeding 121
Conventional Food Chemistry. 122
How the Hen Unbalances Balanced Rations. 126

Chapter 8. Diseases 130
Don't Doctor Chickens. 130
The Causes of Poultry Diseases 131
Chicken-Cholera 133
Roup 134
Chicken-Pox, Gapes, Limber Neck. 135
Lice and Mites. 135

Chapter 9. Poultry Flesh And Poultry Fattening 138
Crate-Fattening 139
Caponizing. 144

Chapter 10. Marketing Poultry Carcasses. 147
Farm-Grown Chickens 147
The Special Poultry Plant 150
Suggestions from Other Countries 151
Cold Storage of Poultry. 153
Drawn or Undrawn Fowls. 154
Poultry Inspection 155

Chapter 11. Quality In Eggs 158
Grading Eggs. 158
How Eggs Are Spoiled 160
The Loss Due to Carelessness. 167
Requisites For the Production of High Grade Eggs 169

Chapter 12. How Eggs Are Marketed. 170
The Country Merchant 170
The Huckster. 172
The Produce Buyer 174
The City Distribution of Eggs 175
Cold Storage of Eggs. 178
Preserving Eggs Out of Cold Storage 181
Improved Methods of Marketing
Farm-Grown Eggs. 183
The High Grade Egg Business 185
Buying Eggs By Weight 188
The Retailing of Eggs by the Producer 189
The Price of Eggs 190

Chapter 13. Breeds of Chickens 197
Breed Tests 197
The Hen's Ancestors. 200
What Breed 204

Chapter 14. Practical and Scientific Breeding 206
Breeding as an Art. 207
Scientific Theories of Breeding. 209
Breeding for Egg Production. 212

Chapter 15. Experiment Station Work **216**

The Story of the "Big Coon" 218

Important Experimental Results at the Illinois Station 222

Experimental Bias 223

The Egg Breeding Work at the Maine Station 225

Chapter 16. Poultry On The General Farm **230**

Best Breeds for the Farm 231

Keep Only Workers 232

Hatching Chicks With Hens 233

Incubators on the Farm 234

Rearing Chicks 236

Feeding Laying Hens 239

Cleanliness 241

Farm Chicken Houses 242

Chapter 1. Is There Money In The Poultry Business?

The chicken business is big. No one knows how big it is and no one can find out. The reason it is hard to find out is because so many people are engaged in it and because the chicken crop is sold, not once a year, but a hundred times a year.

Statistics are guesses. True statistics are the sum of little guesses, but often figures published as statistics are big guesses by a guesser who is big enough to have his guess accepted.

A BIG BUSINESS; GROWING BIGGER

The only real statistics for the poultry crop of the United States are those of the Federal Census. At this writing these statistics are nine years old and somewhat out of date. The value of poultry and eggs in 1899, according to the census figures, was $291,000,000.[*] Is this too big or too little? I don't know. If the reader wishes to know let him imagine the census enumerator asking a farmer the value of the poultry and eggs which he has produced the previous year. Would the farmer's guess be too big or too small?

From these census figures as a base, estimates have been made for later years. The Secretary of Agriculture, or, speaking more accurately, a clerk in the Statistical Bureau of the Department of Agriculture, says the poultry and egg crop for 1907 was over $600,000,000.

The best two sources of information known to the writer by which this estimate may be checked are the receipts of the New

---

[*] The value for 1999 was $22.4 billion, an increase of roughly a hundredfold. Inflation for the same period was only about twentyfold. [Note: All footnotes in this book are by the editor.]

York market and the annual "Value of Poultry and Eggs Sold," as given by the Kansas State Board of Agriculture.

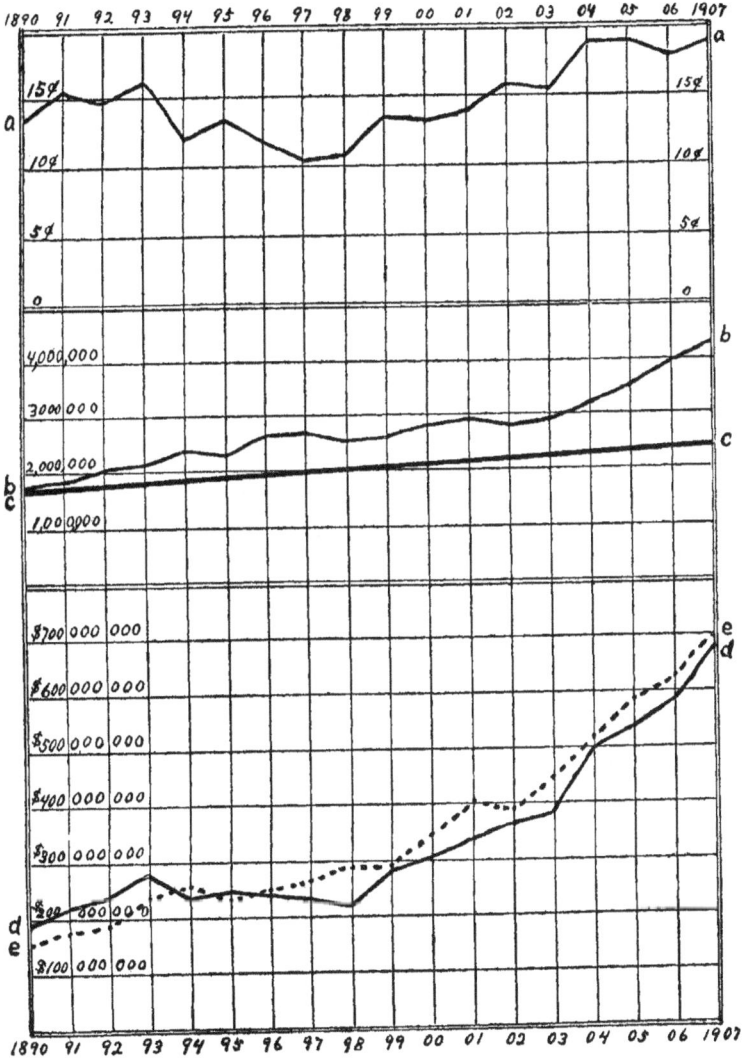

Figure 1. Price, volume, and value of the annual egg crop.

In figure 1 the top curve *a-a* gives the average spring price of Western first eggs[*] in the New York market. The curve *b-b* gives the annual receipts of eggs at New York in millions of cases. Now, since value equals quantity multiplied by price, and since the quantity and values of poultry are closely correlated to those of eggs, the product of these two figures is a fair means of showing the rate of increase in the value of the poultry crop. Starting with the census value of $291,000,000 for the year 1899, we thus find that by 1907 the amount is very close to $700,000,000. This is represented by the lower line.

The value of the poultry and eggs sold in Kansas have increased as follows:

Year	Value
1903	$6,498,856
1904	$7,551,871
1905	$8,541,153
1906	$9,085,896
1907	$10,800,082

The dotted line *e-e* represents the increase in the national poultry and egg crop estimated from the Kansas figures. Evidently the estimate given in Secretary Wilson's report was not excessive.

Now, I want to call the reader's attention to some relations about which there can be no doubt and which are even more significant. The straight line *c-c* in Figure 1 represents the rate of increase of population in this country. The line *b-b* represents the rate of increase in egg receipts at New York. As the country data backs up the New York figures, the conclusion is inevitable that the production of poultry and eggs is increasing much more rapidly than is our population.

---

[*] Roughly Grade B quality.

"Over-production," I hear the pessimist cry, but unfortunately for Friend Pessimist, we have a gauge on the over-production idea that lays all fears to rest. When the supply of any commodity increases faster than the demand, we have overproduction and falling prices. Vice-versa, under-production is shown by a rising price. That prices of poultry and eggs have risen and risen rapidly, has already been shown.

"But prices of all products have risen," says one. Very true, but by statistics with which I will not burden the reader, I find that prices of poultry products have risen more rapidly than the average rise in values of all commodities. This shows that poultry products are really more in demand and more valuable, not apparently so. Moreover, the rise in the price of poultry products has been much more pronounced than the average rise in the price of all food products, which proves the growing demand for poultry and eggs to be a real growing demand, not a turning to poultry products because of the high price of other foods, as is sometimes stated.

LESS HAM AND MORE EGGS

Certainly we, as a nation, are rapidly becoming eaters of hens and of hen fruit. Reasons are not hard to find. Poultry and eggs are the most palatable, most wholesome, most convenient of foods. Our demand for the products of the poultry yard grows because we are learning to like them, and because our prosperity has grown and we can afford them.

Another reason that the consumption of eggs is growing is because the condition in which they reach the consumer is improving. The writer may say some pretty hard things in this work about the condition of poultry and eggs as they are now marketed, but any old-timer in the business will tell you stories of things as they used to be that will easily explain why our fathers ate more ham and less eggs.

Yet another reason why the per capita consumption of hens as measured in pounds or dollars increases, is that the hen herself has increased in size; whereas John when he was Johnnie ate a two-ounce drumstick, now Johnnie eats an analogous piece that weighs three ounces. Perhaps, also, we have a growing respect for the law of Moses, or maybe vegetarians who think that eggs grow on egg plants are becoming more numerous.

Our consumption of pork per capita has, in the last half century, diminished by half, our consumption of beef has remained stationary, but our consumption of poultry and eggs has doubled itself, we know not how many times, for a half century ago the ancestor of the industrious hen of this age serenely scratched up grandmother's geraniums and was unmolested by the statisticians.

WHO GETS THE HEN MONEY?

Seven hundred millions of dollars is a lot of money. Who gets it? There are no Rockefellers or Armours in the hen business. It is the people's business. Why? Because the nature of the business is such that it cannot be centralized. Land and intelligent labor, prompted by the spirit of ownership, is necessary to succeed in the hen business. Land the captains of industry have not monopolized, and labor imbued with the spirit of ownership they cannot monopolize. The chicken business is, in dollars, one of the biggest industries in the country. In numbers of those engaged in it, the chicken business is the biggest industry in the world—I bar none.

Why is this true? Primarily because the hen is a natural part of the equipment of every farm and of many village homes as well. It is these millions of small flocks that count up in dollars and men and give such an immense aggregate.

More than ninety-eight percent of the poultry and eggs of the country are produced on the general farm. The remaining one or two percent are produced on farms or plants where chicken culture is the cash crop or chief business of the farmer. It is this business, relatively small, though actually a matter of millions, that is commonly spoken of as the poultry business, and about which our chief interest centers. A farmer can disregard all knowledge and all progress and still keep chickens, but the man who has no other means of a livelihood must produce chicken products efficiently, or fail altogether—hence the greater interest in this portion of the industry.

The poultry business, as a business to occupy a man's time and earn him a livelihood, is a thing of recent origin and was little heard of before 1890. Since that time it has undergone a somewhat painful but steady growth. Many people have lost money in the business and have given it up in disgust, but on a whole the business has progressed wonderfully, and now shows features of development that are clearly beyond the experimental stage and are undoubtedly here to stay.

The suggestion has been made, by those who have failed or have seen others fail in the poultry business, that success was impossible because of the destructive competition of the farmer, whose expense of production is small. Herein lies a great truth and a great error. The farmer's cost of production is small, much smaller than that on most of the book-made poultry farms—but the inference that the poultryman's cost of production cannot be lowered below that of the farmer is a different statement.

The farm of our grandfather was a very diversified institution. It contained in miniature a woolen mill, a packing house, a cheese factory, perhaps a shoe factory and a blacksmith shop. One by one these industries have been withdrawn from general farm life and established as independent businesses. Likewise, our dairy farms, our fruit farms, and our market gardens have been segregated from the general farm. This simply means that manufacturing cloth or cheese, or producing milk or tomatoes can be done at less cost in separate establishments than upon a general farm.

The general farm will always grow poultry for home consumption, and will always have some surplus to sell. With the surplus the poultryman must compete. His only hope of successful competition is production at lower cost. Can this be done? It is being done, and the numbers of people who are doing it are increasing, but they spend little money at poultry shows, or with the advertisers of poultry papers, and hence are little heard of in the poultry world.

The people whose names and faces are in the poultry papers are frequently there only while their money lasts. They write long articles and show pictures of many houses and yards to prove that there is money in the poultry business, but if one should keep their names and put the question to them five years hence, a great many could say, "Yes, there is money in the poultry business; mine is in it."

Such people and such plants do not get the cost of production down below the farmer's level. Between these two classes of poultry plants, the writer hopes in this work to show the distinction.

Chapter 2. What Branch Of The Poultry Business?

The chicken business is especially prone to failure from a disregard of the common essential relation of cost and selling price necessary to the success of any business. That this should be more true of the poultry business than of any other undertakings is to be explained by the facts that, as a business, it is new; that many of those who engage in it are inexperienced; but most particularly because practically all the literature published on the subject has been written in the interest of those who had something to sell to the poultryman. As a result the figures of production are generally given higher than the facts warrant. The investor, be he ever so shrewd a man, builds upon these promises, and when he finds his production lower, is caught with an excessive investment and a complicated system on his hands, which make all profits impossible and which cannot readily be adapted to the new conditions.

Estimates of poultry profits are quite common, but there are few published figures showing the results that are actually obtained under practical working conditions. In this volume I will try to give the facts of what is being and can be actually accomplished.

Various Poultry Products

In considering the poultry industry we must first get some idea of the various articles produced for sale.

It is common knowledge that the large meat packer can undersell the small packer because the by-products such as bristles, which are wasted by the local killer, are a source of income to the large packer. Now, this does not infer that the small packer is shiftless and neglects to save his bristles, but that on the scale on which he operates it would cost him more to save the bristles than he could realize on them.

So it is with poultry farming. For illustration: A visionary writer in a leading poultry paper, not long ago, advised poultrymen to put spring eggs into cold storage until fall, when prices are high. In reality this would be the height of folly unless the poultryman had his own retail ice plant. In the first place, profit on cold storage eggs, when all expenses are paid, will not average a half a cent a dozen; in the second place, the small lot would be relatively troublesome and expensive to handle, and in the third place, small lots of cold storage eggs are looked upon with suspicion and do not find ready sale. So we see that cold storage eggs are not a suitable product for the small poultryman to handle.

A second illustration of an ill-chosen combination might be taken in the case of a duck farmer who attempts to produce broilers. The principal difficulty of the duck business is that of getting sufficient intelligent labor in the rush season. The chief expense of investment is for incubators and brooder houses. If the duck farmer now tries to add broilers, he will find that the labor comes at the same time of the year, that the chief equipment required is that which is already crowded by the duck business, and that of the men who have succeeded moderately well in caring for ducks will fail altogether with the young chicks, which do not thrive under the same machine-like methods.[*]

On the other hand, let us take the example of an egg farm man who has resolved to combine his attention wholly to the production of market eggs. He succeeds well in his work and is visited by the poultry editors. His picture, the picture of his chickens and of his chicken houses are printed in the poultry papers. For a reasonable sum invested in advertising and in exhibition at the shows, this man could now double his income by going into the breeding stock business. To refuse to spread out in this case would certainly be foolish.

The following classification of the sales products of the poultry industry is given as a basis for farther consideration.

---

[*] Ducks are resistant to most diseases that cause losses in chicks, which allowed the high-density duck industry to predate the high-density broiler industry.

CHICKENS

- For food purposes
  - Eggs
  - Fowls (hens, after laying has finished)
  - Cockerels, necessarily hatched in hatching pullets for layers, (sold as squab broilers, regular broilers, spring chickens, roasters or capons)
  - Both sexes as squab broilers, broilers or roasters.
- For stock purposes
  - Eggs for hatching
  - Day-old chicks
  - Mature fowls

DUCKS

- For table—"green" or spring ducks
- By-products, old ducks and duck feathers
- For breeding-stock

GEESE

- Food, Feathers, Breeders

TURKEYS

- Food, Breeders

PIGEONS

- Squabs, Breeding Stock

GUINEAS

- Broilers, Mature Fowls

I will now discuss these products more in detail. Poultry other than chickens I do not care to discuss at length, because it is not for the purpose of the book, and because the demand for other kinds of poultry is limited and the chance for the growth of the business small.

THE DUCK BUSINESS

The duck business is the most highly commercialized at the present time of any branch of the poultry business. The duck is the oldest domestic bird and was hatched by artificial incubation in China, when our ancestors were gnawing raw bones in the caves of Europe. The duck is the most domestic of birds and will thrive under more machine-like methods and without that touch of nature and of the owner's kindly interest, so necessary to the welfare of the fowls of the gallinaceous order. The green duck business is about twenty years old and has become an established business in every sense of the word. The largest plants now produce about one hundred thousand ducks per annum. The profits at present are not large even for the most successful plants, because the demand is limited and the production has reached such a point that cost of production and selling price bear a definite relation as in all established businesses. The green duck business is not an easy one for the novice because the margin between cost (chiefly food cost) and selling price is low, and unless the new man can reduce the cost of production or raise his selling price in some way, he will have no advantage over the old and successful firms.

SQUAB BUSINESS OVERDONE

The business of producing pigeon squabs resembles the duck business in the sense that it has been reduced to a successful system. The production of squabs has grown until the demand is satisfied and the price has fallen to just that figure that will continue to bring in a sufficient number of squabs from the plants which are already established, or which continue to be established by those who do not stop to investigate the relation between the cost of production and the prevailing prices.

Turkeys Not a Commercial Success

In the case of turkeys, we find exactly opposite conditions. The price of turkeys has risen with the price of chickens and eggs, until one would think that there would be great money in the business, and there is, for the motherly farm wife who has the knack of bringing the little turks through the danger of delicate babyhood. But just as the duck is more domesticated than the chicken, so the turkey, which yet closely resembles its wild ancestor, is less domestic and has as yet failed to surrender to the ways of commercial reasoning, the chief factor of which is artificial brooding.

The presence of a disease called blackhead has done vast injury to the turkey industry in the northeastern section of the country. In the South the industry has been booming. Especially in Tennessee and Texas, I found great local pride in the turkey crop. I certainly would advise any farm wife, in sections where blackhead does not prevail, to try her hand at turkey raising. As to her advisability of continuing in the business, the number of turkeys at the end of the season will be the best judge.

Guinea Growing a New Venture

The guinea growing business is the newest of the poultry industries. In fact, it may be said of guineas, as of our grandmother's tomatoes, "Folks had them around without knowing they were of any use." The new use for guineas is as a substitute for game. Guinea broilers make quail-on-toast and older ones are good for grouse, prairie chicken or pheasant. The retail price in the large cities runs as high as $1.50 to $2.00 a pair. It will probably not pay to raise them unless one is sure of receiving as much as 50¢ each. As for the rearing of guineas, they may be considered on a parallel case with turkeys, if anything they are even more difficult to raise in large quantities. I would also advise this additional precaution: Look up the market in the locality before attempting guinea rearing.

GEESE—THE FAME OF WATERTOWN

As for the goose business, the writer must admit that he doesn't know much about it. In fact, the most of my knowledge concerning this business was acquired by a visit to Watertown, Wis., which is the center of the noodled goose industry.

The Watertown geese are fed by hand every two hours, day and night. They sell to the Jewish trade at as much per pound as the goose weighs, and have brought as high as $14.00 apiece. All of this is interesting, but I hold that the reader who is willing to take instruction will do better to be guided toward those branches of the poultry industry for the products of which there is a great and increasing demand.* So we will leave the goose and guinea business to the venturesome spirits and consider the various branches of the chicken industry.

THE ILL-OMENED BROILER BUSINESS

The broiler business stands today as the ill-omened valley in the poultry landscape. As a rule broiler production has not and probably will not pay. I know of a few exceptions—about enough to prove the rule.†

Most poultry writers, when they make the statement that broilers do not pay, insert the phrase "As an exclusive business" after the word "broilers." This is merely a ruse to take the rough edge off an unpleasant statement, for it certainly hurts the poultry editor to admit that a much-exploited branch of the industry is a failure. Nevertheless, it is a failure, and the more frankly we admit the fact, the less good capital and good brains will be wasted in the

---

* This is very good advice. Exotic products such as fat geese are something to take up as a sideline once you're already established, not something to bet the farm on.

† Hastings' analysis held good for about fifteen years after this book was published. In the mid-Twenties, nutritional advances, in which Hastings himself was involved, turned the broiler business into a commercial success.

attempt to produce at a profit something which is, and probably always will be, produced at a loss.

The reason the broiler is produced at a loss is that 95% of the broilers produced are a by-product of egg, fancy and general poultry production, and as such their selling price is not determined by the cost of production, or the supply determined by the demand. That the broiler business received the boom that it did is due to plain ignorance of the cost of production, or failure to appreciate that the ability to rear young chicks can find a more profitable outlet than in broiler production. Let us take an analogous case. Suppose a city man discovered that there was a demand for dried casein from skim milk. With pencil and paper he could easily figure profits in the business. If this dreamer would attempt to keep cows for the production of casein and throw away his butter fat, we would have an analogous case to the broiler raiser who does not keep his pullets for egg production.

The young cockerel, like skim milk, is a by-product and may pay over the cost of feeding, or some other specific item—but that he does not pay the whole cost, including wages for the manager is proven by two facts:

1. Every large broiler plant yet started has either failed flatly or shifted its main line to other things.
2. Egg farmers would be only too glad to buy pullets at the price for which they sell the cockerels—a confession that it costs more to produce broilers than they will bring.

The conception of the broiler business when it boomed twenty years ago was to produce broilers in early spring when other folks bad none. It was, like the early watermelon or the early strawberry business—to make its profit in extreme prices.

This idea received several severe blows from the hands of modern progress. One is the development of poultry fattening and crate-feeding in this country. This has resulted in supplying the consumer with choice chicken-flesh that can he produced more economically than broilers. Formerly it was a case of eating old hen, rooster (age unknown), or broilers—now we have capon, roaster, crate-fattened chickens and green ducks, all rivals for the place formerly occupied exclusively by the broiler.

Again, the improvement of shipping and dressing facilities, the universal introduction of the refrigerator car and the introduction into the central west of the American breeds has flooded the eastern market with a large amount of spring chickens—by-products of the egg business on the farm—which are almost equal in quality to the down-eastern product.

The most prominent reason of the lessened profit in broilers is the development of the cold storage industry. Cold storage destroys the element of season, and allows only that margin of profit that the consumer willing to pay for a fresh killed broiler from a Jersey broiler plant as compared with last summer's product from the Iowa farms. From a summer copy of *Farm Poultry,* I quote the Boston market:

Fresh killed Northern and Eastern:

Fowls, choice:	15¢
Broilers, choice to fancy:	23¢-25¢

Western, Ice Packed:

Fowls, choice:	14¢
Broilers, choice:	20¢-22¢

Western Frozen:

Fowls, choice:	14¢
Broilers, choice:	18¢-20¢

Eggs:

Nearly fancy:	26¢
Western choice:	$17\frac{1}{4}$¢-$18\frac{1}{2}$¢

To complete our comparison I turn to the previous winter and find that the best storage eggs are quoted at 19¢, when the best fresh are selling at 35¢. This was a poor storage season and a quotation of 22¢ and 25¢ would perhaps be a fairer comparative figure. We find the percent of premium on the local product to be:

Fowls, local over fresh western:	7%
Fowls, local over frozen western:	7%
Broilers, local over fresh western:	14%
Broilers, local over frozen western:	26%
Eggs, local over fresh western:	30%
Eggs, local over storage western:	37%

I consider these general facts concerning the failure of broiler production, and the logical explanations given, as far more convincing than any figures I could give concerning the detailed cost of production. Nor am I capable of giving as accurate figures as I can in the case of poultrykeeping for egg production, for I have had neither the desire nor the opportunity to look them up.

The following suggestive analysis I submit for the purpose of pointing out why the cost of production is too great to allow a profit. Consider the early broiler going to market in May: its weight at 1¼ pounds and the price at 35¢ a pound, or, putting it roundly, a price of 50¢ a bird.

Now, May broilers mean February eggs. If the reader will refer to the tables of hatchability and mortality he will see that, for our northern states, this is one of the worst seasons for hatching. A hatchability of 40% times a liveability of 50% gives a net liveability of 20%. Now, anyone with the ability to produce high grade eggs at that time a year, could get about 40¢ a dozen for them, which raises the egg cost per broiler to about 17¢. The feed cost per broiler is small, usually estimated at 12¢, and this makes a cost of 29¢. Now, let us allow a cent for expense of selling charges and forget all about investment, fuel and incidentals, we have left a margin of 20¢.

Before going farther let us look at the labor bill. Suppose it is a one-man plant. Suppose the owner sets a value on his services of $1,200 per annum. That is pretty good, but few men who set a lower value on their services will have accumulated enough capital to go into the business. At 20¢ each, it will take 6,000 broilers to make $1,200. That will take 30,000 eggs and at three settings will require forty 240-egg incubators, which, of a good make, will cost $1,260. To spread the hatching out over a longer period is to run into cheap prices on the one hand, or a still-impossible egg season on the other. It will take upwards of a hundred brooders to house the chicks.

There is no use of going farther till we have solved these difficulties. First, we have more work than one man can do; second, we require a number of hatchable eggs that cannot be bought in winter without a campaign of advertising and canvassing for them that would make them cost double our previous figure. To produce them oneself would require a flock of 2,500 hens. When a man gets to that point, he is out of the broiler business and has become an egg farmer, and will do the same thing as every other egg farmer—hatch the chicks when eggs are cheap and fertile, selling his surplus cockerels for 25¢ each, and permitting the cold-storage man to freeze them until the following spring to compete with the broiler man's expensively produced goods.

The effort at early broiler production was a natural result of the combination of the idea of artificial incubation with our grandmother's pride in having the first setting hen. But, in the present age, the man who attempts it is rowing against the current of economical production, for the cheaply produced broiler can be stored until the season of scarcity, with but slight loss in quality. To produce broilers in the season of scarcity necessitates the consumption of another out-of-season product (eggs), which cannot be so successfully stored. We will give the production of broilers no further attention save as a by-product of egg production.[*]

---

[*] In spite of this apparent dismissal, Chapter IX and Chapter X are given over to the production and marketing of poultry meat.

SOUTH SHORE ROASTER

The production of South Shore soft roasters in a local section of Massachusetts offers a successful contrast with the broiler business and is, so far as the writer knows, the only case in the United States where pullets are profitably diverted from egg production.

The process of master production is essentially as follows: The incubators are set in the fall or early winter, and the chicks reared in brooder houses. As soon as the tender age is past, the chickens are put in simple colony houses (portable sheds) where, with hopper-fed corn, beef scrap and rye on the range, they grow throughout the winter and spring. They are sold from May 1st to July 1st and bring such prices that the cockerels are caponized yet not sold as capons, showing them to be the highest priced chicken flesh in the market, except for small broilers. Now, the income of roasters is two to five times as much per head as that of broilers. The added expense is only a matter of feed, which bears about the same ratio to weight as with broilers. The great advantage of the roaster business over that of the broiler business comes in the following points:

1st: The initial expense of eggs, incubation and brooding are distributed over a much larger final valuation.

2nd: The incubation period, while perhaps in as difficult a season, can be distributed over a longer period of time.

With eight-pound roasters at 30¢ per pound, we have an expense account about as follows: cost of production to broiler stage, 30¢ as previously given. An additional food cost of 10¢ per pound of chicken flesh would still leave a margin of $1.40; so, for an income of $1,200, only about 860 birds need be raised, a proposition not beyond the capacity of one man to handle.

Allowing a spread of five hatching periods, the number of eggs required at any one time would be one-twelfth that demanded by the broiler farm. As it is, the roaster grower finds trouble in getting good eggs and is obliged to pay 50¢ a dozen for them, but this is within the region of possibility.

The South Shore roaster district is an example of an industry built up by specialization and cooperation. But in this sense I do not mean cooperation in production, but that the product is handled by a few dealers and has become well-known so that the brand sells readily at an advanced price. To a beginner in the South Shore district, the numerous successes and failures around him cannot help but be of great benefit. The South Shore roaster district of Massachusetts is the best example of specialized community production of poultry flesh that we have in the United States. It is only rivaled by the districts in the south of England and in France.

In Chapter III the writer takes up fully the community production of eggs. The reason I have gone into this matter in regard to eggs, rather than roasters, is because the egg production is much the greater industry, and, whereas the soft roaster is at a premium only in a few Boston shops, high-grade eggs are universally recognized and in demand. Many of the economies, especially concerning incubation, would apply equally well in both communities. I expect to see the time when chicken flesh shall be produced with these more advanced methods in many "South Shore" communities.

TOO MUCH COMPETITION IN FANCY POULTRY

The various types of chicken farming are classified by what is made the leading sales product. This will depend wholly upon what is done with the female chicks that are hatched. If they are sold as broilers it is a broiler plant; if as roasters, it is a roaster plant; if as stock, it is a fancy or breeding stock business, but if kept for laying the proposition is an egg farm, and all other products are by-products. These by-products are to be carefully considered and sold at the greatest possible price, but their production is incidental to the production of the main crop.

Of the fancy poultry business as a main issue, it must be said that it is certainly a poor policy to start out to make a living doing what hundreds of other people are only too glad to spend money in doing. Just as a homeless girl in a great city is beaten out in the struggle for existence by competition with girls who have good homes and are working for chocolate money, so the man starting out as a poultry fancier is certainly working at great odds in competition with the professional men, farmers and poultry raisers whose income from fancy stock is meant to buy Christmas presents and not to pay grocery bills.

To enter the fancy poultry business, one should take up poultry breeding in a small way, while working at another occupation, or he may take up commercial poultry production and learn to produce stock in large quantities and at a low cost, after which any breeding stock business he may secure will be added profit. The fancier will find the cost of production as given for commercial purposes very instructive, but if he operates in a small way he should expect to find his productive costs increased unless he chooses to count his own labor as of little or no value. That every chicken fancier also has, in a small way, commercial products to sell goes without saying. These, indeed, together with his sales of high-priced stock, may pull him through with a total profit, even though his production cost is great, but every fancier should take a pride in making the sales at commercial rates pay for their cost of production.

If the reader has received the impression from the present discussion that fancy poultry breeding always proves unprofitable, he certainly has failed to get the key-note of the situation. There are numbers of fancy poultry breeders making incomes of several thousand dollars per year, but they are old breeders and well-known men.

There is another type of poultry fancier who is more commercial in his methods, but whose work lacks the personal enthusiasm and artistic touch of the regular fancier. I refer to the bandwagon style of breeder who gets out a general catalog in which are pictured acres of poultry yards with fences as straight as the draftsman's rule can make them. Such men do a big business. They may carry a part or all of the breeding stock on a central poultry plant and farm out the eggs, contracting to buy back the stock in the fall, or the poultry farm may be a myth and the manager may simply sell the product of the neighboring farmers who raise it under contract.

The system is naturally disliked by the higher class fanciers, but the writer must confess that any system which gets improved stock distributed among the farmers is worthy of praise. These types of poultry farms have been more largely carried on in the West than in the East, owing to the fact that true fanciers are thicker in the East. There is undoubtedly still plenty of room for bandwagon poultry plants in the West and especially in the South.

As adjuncts of this business may be mentioned the sale of a line of poultry supplies and the handling of other pet stock, such as dogs or Shetland ponies. In this case the advantage of such additions depends upon the fact that the greatest cost is that of advertising, and, if anything that will be associated in the buyer's mind with the main article is added to the catalog, it will result in additional sales at a low rate of advertising cost.

EGG FARMING THE MOST CERTAIN AND PROFITABLE

We have now discussed all the branches of the poultry business except that of egg production, and the result of our review indicates that most of these fields provide either limited opportunities or present obstacles in the very nature of the work that prevent their being conducted on a large scale.

Egg production is undoubtedly the most promising and profitable branch of the poultry industry. The chief reason that this is true is to be found in the fact that the most difficult feature in chicken growing is the rearing of young stock through the brooding period. Now, as the eggs laid by a hen are worth several times the value of her carcass, it stands to reason that once we succeed in rearing pullets, egg farming must be the most profitable business to engage in.

For each hen that passes through a laying period there is her own carcass, and at least one cockerel that are necessarily produced and that must be marketed. Now, the pullet is worth more for egg producing than can be realized for her as a broiler or roaster, and her extra worth may be considered as counter-balancing the price at which cockerels must be sold.

The egg crop represents about two-thirds of the value of all poultry products, and the demand for the high-grade goods has never been satisfied. Egg farming cannot easily be overdone, whereas any other type of poultry production must compete with the cockerels and hens that are a by-product of egg farming.

Egg farming by no means relieves one from the difficulties of incubation and growing young stock, but it does throw these difficult parts of the business at the natural season of the year and results in a distribution of work throughout a longer period of time.

In the remainder of the volume we will consider the poultryman as an egg farmer. We will also, unless otherwise stated, assume that he is a White Leghorn egg farmer, who is hatching by artificial incubation. Such reference to the marketing of poultry flesh or to other breeds will be made only in comparison of this type of the business or in relation to the production or handling of farm-grown poultry.[*]

---

[*] By 1999, eggs had fallen to less than one-fourth of the total value of the poultry crop. However, the demand for high-grade products still hasn't been satisfied. The egg business is easier to learn than the broiler business, though the high-grade broiler business probably returns a greater hourly income and return on investment than the high-grade egg business once you're fully up to speed.

Chapter 3. The Poultry Producing Community

The builder of air castles in Poultrydom invariably starts out with a resume of the specialization of the world's work and the wonderful advances in the economy of production of the large corporate organizations, compared with the individual producer.

The lone blacksmith hammering out a horseshoe nail is contrasted with the mills of the American Steel Company. The fond dreamer looks upon the steel trust, the oil trust, the department store, the packing house, the chain groceries, the theatrical trust, and the colossal enterprises that dominate every field of industry save agriculture. Here, then, lies the neglected opportunity for the industrial dreamer to hop over the fence, awaken the sleeping farmer, and fill his own purse with the wealth to be made by applying modern business methods to agriculture.

The knowing smile—the farmer may be asleep and he may not be. Suppose that he is: does the fond dreamer dream that he is the first man from the industrial kingdom of great things to look with hungry eyes at the rich field of agricultural opportunity, basking in last century's sun? Alas, fond dreamer, your name is legion. Every farmer who has sent a son to college has known you and the Hon. William Jennings Bryan has met you, called you an agriculturist, and defined you as a man who makes his money in town and spends it in the country.

But the dreamer is right in his first premise—great economies in production are the result of specialization and combination. Why not then in agriculture? I'll tell you why. There is a touch of nature in the living thing that calls for a closer interest on the part of the laborer than the industrial system of the mine and factory can give.

Why is combined and specialized production more economical? It may be because it gets more efficient work out of labor, it may be that larger operations make feasible the employment of more efficient methods and machinery. The cost of production may be lowered by either or both of these means, or it may be lowered by an increased efficiency in machinery, even with a decreased efficiency in labor.

Combination and specialization so commonly cut down expenses because of large operations and the use of better tools, that we may take this saving for granted. When it comes to labor there is a different story. The manual laborer working with boss and gang or the machine-tender in the factory work as well or better for large than for small concerns, but the labor of a poultry plant is different. It is made up of a great many different operations well scattered in space and time. For the most part it is simple labor, but it is essential that it be performed with reasonable concern for the welfare of the business.

In other industries, as with men working at a bench, the presence of a foreman keeps them busy and their work may be daily inspected. To have foremen in poultry work would require as many foremen as laborers, and even then they would be as useless, for when the last round of the brooders is made at night a foreman standing three feet away could not knew whether the laborer who had placed his hand in the brooder had found all well or all wrong.

It is useless to carry the argument farther. The labor bill is one of the biggest items of expense in poultry production. With a system where the efficiency of the labor decreases with the size of the business, large industrial enterprises are impossible. Such savings as will be made in buying supplies, selling, etc., will be wasted in the reduced efficiency of labor.

The bulk of labor in poultry work must be self-reliant labor and the only test for such efficiency is number of chicks reared and the weight of the egg basket. Even this will not he a complete test unless from the income is subtracted the feed bills.

A system of renting or working on shares, that will gain the advantages of centralization without losing the individual interest of the laborer, will go a long way toward making the poultry business one wherein large capital and large brains can find a place to work. I expect to see in the future some such system evolved. In fact we have today a profit-sharing plan between owner and fore-man on many of our best plants.[*]

To extend such to each laborer requires more system and bet-ter superintendence, but it is feasible and must come. But, better still is it for the worker to own the stock. Best yet if he owns both stock and land, leaving to larger capital only such phases of the business as involve great saving when done on a wholesale basis.

Just as the manufacturer of farm machinery, the packing of meat and the manufacture of butter have successfully been taken out of the control of the individual farmer and placed under cor-porate or cooperative organization, so the writer expects to see certain portions of the process of poultry production removed from the hands of the farmer and controlled by more specialized and expert labor. Far from meaning the lessening of the earning power of the farmer, every one of such steps means larger pro-duction and more profits. The ideal of agricultural economics is to give the farmer the smallest possible proportion of the work of agricultural production in order that the most may be produced and the farmers share along with the others may be largest.

ESTABLISHED POULTRY COMMUNITIES

In a previous chapter we spoke of the South Shore roaster district of Massachusetts. Here is a community where, in lots of from a dozen to four or five thousand, are annually produced seventy-five to one hundred thousand market fowls of one particular type. While this business was not built up by the efforts of a corpora-tion or individual who planned definitely the entire project, yet we find a central influence at work in the person of the firm of Curtis

---

[*] The modern broiler industry works pretty much along these lines, though I doubt that many people are entirely happy with the results. As I write this (March, 2003), nei-ther the contract growers nor the giant poultry companies are making much money. In the egg industry, the invention of the laying cage and the resulting high level of automa-tion made it possible to run the business with hired labor.

Bros., who for years have bought the majority of South Shore roasters, and who have done a great deal to advertise the product and encourage their neighbors to a larger and more uniform production.

At Little Compton, R. I., is a very similar parallel of the South Shore district in the shape of egg farms. Here we find within a radius of two miles about one hundred thousand Rhode Island Red hens owned in flocks of two thousand or less. The methods used throughout the community are all alike and are simple in the extreme. There are no incubators, no brooders, no poultry houses, no long houses, no dropping hoards to be scraped every morning; nothing, in fact, but board-walled, board-roofed colony houses, scattered over the grass fields—and similar though smaller fields covered with coops for hens and their chicks. Feeding is equally simple; a mash of meat, vegetables and ground grain mixed once a day and hauled around in a one-horse cart, and hoppers of whole corn exposed in the houses. The houses are cleaned twice a year. Little Compton is, indeed, a community where all the rules of the poultry books are regularly violated, and yet a larger number of successful egg farms can be seen from the church spire at Little Compton Corners than most poultry writers have ever seen or read about. Strange it is, as Josh Billings puts it, that "some folks know things that ain't so."

An illustration published in a recent issue of the *World's Work* tells a remarkable story. A pile of egg shells as big as a straw stack certainly indicates "something doing" in the chicken business, and it is a very proud monument to Mr. Byce who, some twenty-odd years ago, established an incubator factory at the town of Petaluma. Petaluma is in Sonoma County, California, forty miles north of San Francisco. In the census year of 1899, Sonoma County produced more eggs than any other county in the United States. Today there are in the Petaluma region close to one million hens.

Like the Little Compton district, Petaluma is a one-breed community, White Leghorns being the breed used. The individual flocks are larger than at Little Compton, chiefly because the milder climate, smaller breed, and establishment of the central hatchery enable one man to take care of more birds.

When I asked Mr. Byce for a list of the people in his neighborhood keeping over one thousand hens, he replied by sending me a list of twenty-two men who keep from 8,000 down to 2,500 each, and said that to give those keeping from one to two thousand would practically be to take a census of the county. The methods of housing and feeding used are simple and inexpensive, like those at Little Compton.

The chief reason why Petaluma shows a more advanced development in the poultry community than the eastern poultry growing localities is to be found in the climatic advantages which favor incubation (see Chapter on Incubation) and the consequent development of the central hatchery. Outside of this, the location is not especially favorable. The temperature is milder in the winter than in the East, but the Petaluma winter is one of continual rain which develops roup[*] to a greater extent than we have it in the East. The prices received for high grade eggs in San Francisco in the winter is about equal to the top prices in New York. In the spring and summer New York will give more for fancy goods. The cost of corn on the Pacific Coast is about 40¢ a hundred more than on the Atlantic Coast. Wheat, however, is cheaper than in the East, but not cheap enough to substitute for the more staple grain.

The eggs from the Petaluma region are at present marketed largely through a cooperative marketing association.

Developing Poultry Communities

I have shown why the large individual poultry farms with hired labor have not proven profitable fields for the investment of capital. Again, I have shown that in a few localities where the business was incidentally started, communities of independent poultry farmers have grown up which are very successful, and that there is no apparent reason why similar communities elsewhere, if intelligently located, could not do as well or better.

This looks like an excellent field for group enterprise. Certainly there is no reason why the poultry community cannot be as

---

[*] An upper respiratory infection, now rare.

successfully promoted as an irrigation project, a cheese factory, or a truck-farming community. In such a community there are many functions that can be better performed by a capitalized body managed by experts than by individual poultrymen acting alone.

These functions are:

1. The selection of a location and the purchase of the land in large quantities.
2. Laying out this land into suitable individual holdings with regard to economy of water supply and the collection of the product.
3. The partial or complete equipment of these farms at less expense and in a more suitable manner than could or would be done by the individual holders.
4. The sale or rent of these places to poultrymen at a reasonable profit on the investment, but at a rate which will still be below the cost at which the individual could have acquired the land.
5. The selection of the stock that would not only be better adapted to the enterprise than that which would be acquired by the individual farmer, and would possess the uniformity necessary to maintain a standard grade in the product.
6. The centralized hatching of the chicks by which means chicks can be more cheaply hatched and better hatched than by the imperfect methods available to the small poultry man.
7. The purchase of all outside supplies with the usual savings involved in large purchases.
8. A system of delivering such supplies to the farm.
9. A general protection against thieves and predatory animals by an organized war on all "varmints."
10. Maintenance of the best methods in feeding and care by the employment of skilled advisers, or the operation of demonstration farms under the direction of the central management.
11. The enforced daily gathering of all eggs and their lodgment that same evening in a clean, dry cooler, with a thermometer hovering around 29° Fahrenheit.[*]

---

[*] I think that 40-45° F is better for short-term storage.

12. The strict enforcement of penalties against the man who attempts to sell bad eggs.
13. The prompt dispatch of the product to its final market.
14. The final sale of the eggs with opportunities for fancy prices made possible by an absolutely guaranteed product in quantities sufficient to permit of a regular supply and of advertising the product.
15. The conduction of breeding operations along any desired line, with the opportunity of combining the principle of great numbers for selection with the comparison of all progeny from ancestry, a method that will bring results a hundred times more quickly than the efforts of the small breeder.
16. The advantage of the sale of breeding stock to be acquired from the free publicity which is showered on all unique industrial enterprises.

In these sixteen functions there is ample opportunity for capital, backed by ability in organization, to reap an ample reward. Is it a dream? In a sense yes, but a dream made possible by the observation of the actual results achieved in similar lines, and of the present tendency in the poultry producing world.

Why has not this thing been done before? Because no one knew enough to do it. Why did not the wonderful truck-farming regions develop earlier in the South, and why does it still take northern settlers, backed by railroad advertising, to develop the wonderful modern industries which enables every city dweller in the North to have strawberries in February and fresh vegetables any day in the year?

Why did the California fruit trade develop? Did anyone suppose forty years ago that the unsettled valley around Pasadena would ever produce one thousand dollars per acre in one year? These orange groves, too valuable for agricultural purposes to be used as town sites, were precarious experiments until the transcontinental refrigerator car and the California Fruit Growers' Exchange paved the way and put each day in every eastern and northern city just the quantity of oranges that the people could consume at a profitable price.

Mr. Harwood, in the *World's Work* for May, 1908, after describing the "City of a Million Hens," raises the question, "If in Petaluma, why not anywhere?" I would like to answer that question by saying that, while anywhere is a little broad, the reason the industry has not developed elsewhere has been because of the diversion of interested capital towards impractically large individual poultry plants manned by hired labor. Another reason has been the lack of the technical knowledge necessary to construct and operate efficient hatcheries.

The poultryman has been a disciple of the poultry papers and poultry fanciers of the day. The poultry papers and poultry literature has generally been supported by poultry fanciers and manufacturers of incubators, patent nests and portable houses. These good folks have vied with one another in complicating the business. They have built steam-piped houses with padded walls and miniature railways with which daily to haul away the droppings. A few famous fanciers selling eggs at $10.00 per setting have made such business pay, but alas for the luckless investor in what the visiting poultry editor would style a "handsomely equipped modem poultry plant."

A few years ago a Government poultry expert paid a visit to Petaluma. He came back and reported, "It is a great disappointment, the methods are very crude." The case is most pathetic. Here was a man employed by the people to teach them how to make poultry pay. His carfare is paid across the continent that he might visit the only community in the United States where at that time any considerable number of people were making their living from poultry, and because he did not find lightning rods on the poultry houses, he came back with the look of Naamen who, when he was requested by Elisha to bathe seven times in the river Jordan, replied, "It is very crude."[*]

---

[*] *2 Kings 5:1-16*

WILL COOPERATION WORK?

That magic thing, "Cooperation," while utterly lacking in the Utopian qualities with which the word artist paints it, is a bigger factor in American affairs than the average man realizes.

The chief difficulty with cooperation is that the manager, if not incompetent, has an excellent opportunity to be a grafter. In Europe cooperation in agricultural and mercantile enterprise is older and better developed than in this country. Perhaps the Europeans are less inclined to be grafters, but a more likely explanation is that the members of such associations as these have learned how to prevent and detect graft, just as our business men have learned to avoid losses from the dishonesty of employes. That this is the true explanation is substantiated by the fact that when cooperation once becomes established in this country, it succeeds even better than in Europe.

When the creameries were started in the West several years ago, there was much complaint of swindlers, fake stock companies, and cooperative ventures in which the manager absconded with the butter money. Today more than half of the American creameries are cooperative and the number is constantly increasing. They are efficient and successful in every way, and today make the finest of butter and pay the highest prices to the farmer for his cream. But their way was first paved and the business developed by successful private concerns.

Cooperation is entirely feasible and successful where the people behind the movement have enough interest in the enterprise and good enough business sense to run the proposition as efficiently as similar private enterprises are run. The idea that cooperation must always result in a big saving is a misconception. Employes will not work any harder for an association than for a private employer, sometimes not as hard. Certainly no employe will work as hard for an association as he will for himself.

Why people should expect to buy out the grocery store and hire the grocer to run it, and save money for themselves, is a thing I could never understand. But if there is some great waste that cooperation will prevent, as where seven milk wagons drive every morning over the same route, or where the market of perishable crops is glutted one day and starved the next, centralization, corporate or cooperate, will pay.

I know of no better way to impress the reader with American cooperation in actual practice than to quote from a brief account of the California Fruit Growers' Exchange.

The Exchange was founded upon the theory that every member is entitled to furnish his *pro rata* of the fruit for shipment through his association, and every association to its *pro rata* to the various markets of the country. This theory reduced to practice gives every grower his fair share, and the average price of all markets throughout the season.

Another cardinal provision of the plan was that all fruit should be marketed on a level basis of actual cost, with all books and accounts open for inspection at the pleasure of the members. These broad principles of full cooperation constitute the basis of the Exchange movement.

The Exchange system is simple, but quite democratic. The local association consists of a number of growers contiguously situated, who unite themselves for the purpose of preparing their fruit for market on a cooperative basis. They establish their own brands, make such rules as they may agree upon for grading, packing and pooling their fruit. Usually these associations own thoroughly equipped packing houses.

All members are given a like privilege to pick and deliver fruit to the packing house, where it is weighed in and properly receipted for. Every grower's fruit is separated into different grades, according to quality, and usually thereafter it goes into the common pool, and in due course takes its percentage of the returns according to grade.

Any given brand is the exclusive property of the local Association using it, and the fruit under this brand is always packed in the same locality, and therefore of uniform quality. This is of great advantage in marketing, as the trade soon learns that the pack is reliable.

There are more than eighty associations, covering every citrus fruit district in California, and packing nearly two hundred reliable and guaranteed brands of oranges and lemons,

The several local Exchanges designate one man each from their membership as their representative, and he is elected a director of the California Fruit Growers' Exchange. By this method the policy-making and governing power of the organization remains in the hands of the local Exchanges.

From top to bottom the organization is planned, dominated and in general detail controlled absolutely by fruit growers, and for the common good of all members. No corporation nor individual reaps from it either dividends or private gain.

So far we have dealt almost exclusively with the organization of the Exchange, its cooperative aspects, and general policy at home. Equally important is its organization in the markets.

Seeking to free itself from the shifting influence of speculative trading, by taking the business out of the hands of middlemen at home, the Exchange found it quite as important to maintain the control of its own affairs in the markets.

For this purpose the Exchange established a system of exclusive agencies in all the principal cities of the country; employing as agents active, capable young men of experience in the fruit business. Most of these agents are salaried and have no other business of any kind to engage their attention, and none of the Exchange representatives handle any other citrus fruits. These agents sell to smaller cities contiguous to their headquarters, or in the territory covered by their districts.

Over all these agencies are two general or traveling agents with authority to supervise and check up the various offices. These general agents maintain in their offices at Chicago and Omaha a complete bureau of information, through which all agents receive every day detailed information as to sales of Exchange fruit in other markets the previous day. Possessing this data, the selling agent cannot be taken advantage of as to prices. If any agent finds his market sluggish, and is unable to sell at the average prices prevailing elsewhere, he promptly advises the head office in Los

Angeles, and sufficient fruit is diverted from his market to relieve it and restore prices to normal level.

Through these agencies of its own the Exchange is able to get and transmit to its members the most trustworthy information regarding market conditions, visible supplies, etc. This system affords a maximum of good service at a minimum cost. The volume of the business is so large that a most thorough equipment is maintained at much less cost to growers than any other selling agency can offer.

The annual business of the California Fruit Growers' Exchange amounts to over ten million dollars, and the Exchange handles over half the citrus fruit output of the State. Yet there are people who say cooperation in America will not work.

COOPERATIVE EGG MARKETING IN DENMARK

I have discussed at length the work of the California Fruit Growers' Exchange, as the best example in the United States of the cooperative marketing of farm produce. We have thus far but little cooperative work in the marketing of poultry products. Canada has a few examples, but it is to European countries that we must go for a full demonstration of the principle of cooperation when applied to the products of the hen. In England and in Ireland cooperative efforts in the growing, fattening, and marketing of poultry and eggs are quite common. It is to Denmark, however, that we must go to find the most wholesale example of this truly modem type of business effort.

The Danes are cooperators in the fullest sense. They have cooperative creameries and cooperative packing houses. The Danish Egg Export Society is an organization, the plan and work of which is very much like that of the California Fruit Growers' Exchange.

The local branch of the association buys the eggs of the farmer, paying for them by weight. Collectors are hired to gather them at frequent and regular intervals, and are paid in accordance with the amount of their collections, but must stand the loss of breakage. Each individual poultryman's eggs are kept separate until they reach a centralizing station. There are a number of these central stations at which the eggs are carefully graded and packed for shipment to England.

The individual farmer is fined or taxed for all bad eggs found in his lot. This fine is deducted from his receipts and he has nothing to do but to submit to it or got out of the association. The latter he cannot afford to do because the association has its established brands and can pay him more for his eggs than he could secure by attempting to market them himself. As a result of this strict system of making the producer responsible for weight and quality of the eggs the Danish eggs have become the largest and finest in the world.

Although the writer firmly believes in the cooperative marketing of farm produce, and considers that the success already secured in this work is conclusive evidence of the practicability and desirability of cooperation, it would not be fair to infer that cooperation has entirely driven out private or corporate enterprise. Just as a goodly percent of the citrus fruit of California is still handled by private dealers, so in Denmark we find that nearly one-half of the eggs sent to England are handled by private companies. Let it be noted, however, that these companies maintain a system of buying on merit which enforces high quality that is not to be found where private buyers are without the spur of cooperative competition. Before cooperation entered the orange regions of California, the fruit was poorly packed and handled and the markets at times so glutted, that shipments of fruit sometimes failed to pay the freight, and this was actually charged back to the unfortunate grower. Cooperation has done away with this waste. In like manner our great loss from decomposed eggs and half hatched chicks is unknown to the egg trade of Denmark.

CORPORATION OR COOPERATION?

The community of farmers producing a large quantity of a single kind of product is the coming form of agricultural enterprise. Will this community be promoted by corporation or by cooperation?

Arthur Brisbane says, "As individual control of the Government has been superseded by collective control, so individual control of industries will be followed by collective control. That is the natural order."

Brisbane is right. The individual, or the corporation, which is an individual using other men's money, foreruns cooperation, because the individual knows exactly what he wants to do and the big group of individuals does not know what they want or how to do it until individuals have, by concrete successes, shown them.

When the creameries were started, cooperative creameries were unsuccessful and could not compete with privately owned creameries. The farmers have now become too wise to be "easy marks" to the fake creamery promoters or to trust their butter sales to a comparative stranger, and cooperation is a success.

Poultry communities cannot be made out of whole cloth by the cooperative plan. Private corporations will be necessary to launch these enterprises. When they have reached the stage of development now to be seen in Little Compton and Petaluma they are ready for cooperation.

I have emphasized the point that the private corporation is the natural forerunner in this matter in order to discourage premature or over-ambitious efforts at cooperation. Whenever a community of poultrymen or, for that matter, a community of growers of any perishable form of products, who are already successful in the producing end, wish to take up cooperation and will see that men are selected to manage it who will use the same precautions to guard against incompetency or graft that they, as individuals, would use in their own business, there is excellent chance of success.

Go slow. Do not expect to get rich quick by "cutting out the middleman's enormous profits," for the middleman's profits are not enormous, and if you see that your cooperation is not paying, give it up and confess to yourselves that you do not know as much about the business as your private competitors.

Chapter 4. Where To Locate

That poultry should be kept on every farm, to supply the farmer's own table, does not permit of argument. When it comes to production for market, I believe there are some sections where it costs more to produce and market poultry and eggs than is received for the product when sold. For example, on a farm which is twenty miles from town and where grain cannot be profitably grown, the cost of hauling grain from the railroad station and of sending the eggs to market as frequently as is necessary to have a wholesome product, would certainly eat up all possible profits.

The farmer thus located would find a more profitable use for his time in some industry where the raw material is near at hand and the product needs less frequent marketing.

SOME POULTRY GEOGRAPHY

When we are discussing poultry on the general farm, the problem of location is not to be taken into consideration, save to the extent that there are a few localities where food cost is so high or marketing facilities so poor as to make even the usual farm surplus unprofitable.

The map on page 51 shows the intensity of egg production and also indicates the location of the more important localities where poultry plants have succeeded. The map on page 53 shows the quality of eggs coming from various sections, which indicates pretty closely the general development of the poultry industry. These indications, however, are of little value in locating a poultry plant, for they refer to the poultry product on the general farm, and are a matter of the number and general intelligence of farmers, rather that a sign of the suitability of the locality for the poultry industry.

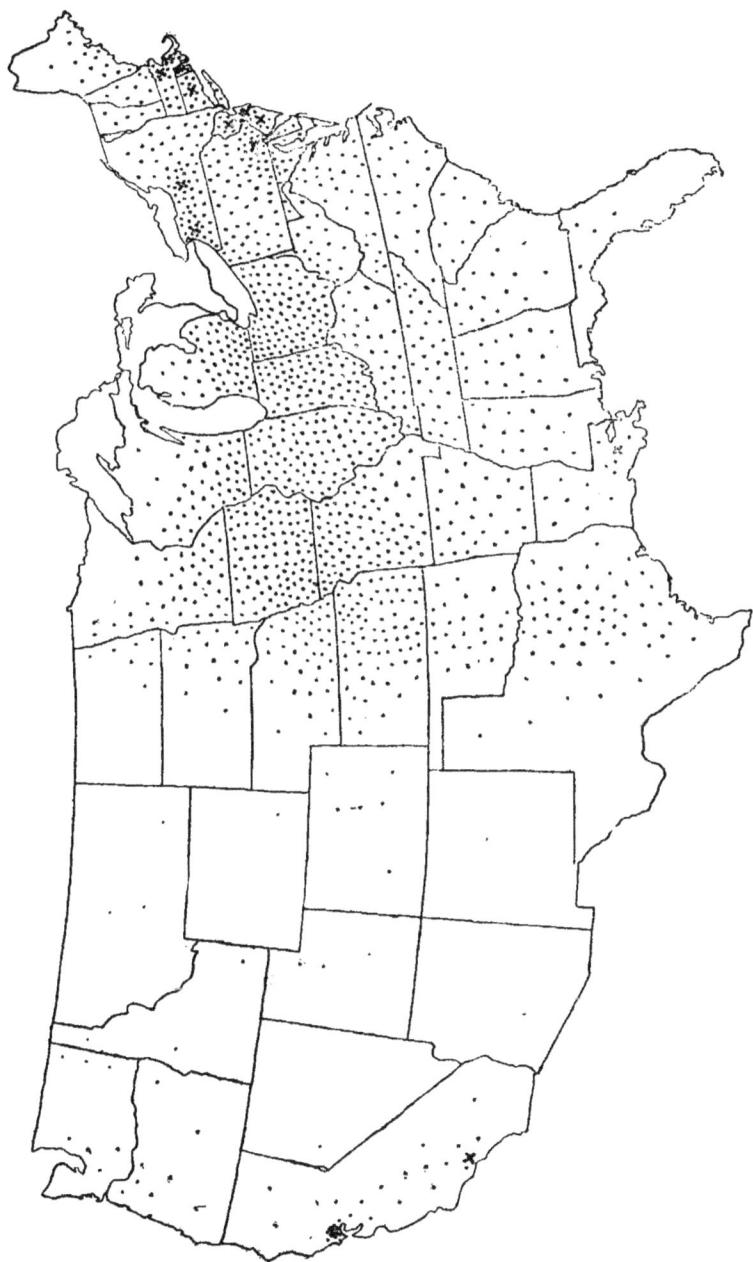

Figure 2. Intensity of poultry farming across the country

For purposes of discussion, I have divided the United States into seven sections as shown by the dotted lines on the second map.

Section 1 is the North Woods and too cold and remote for the poultry business,

Section 2 includes the great West, of which an adequate discussion is out of the question. Of course, the great majority of this area is too remote from markets for poultry production. The locations around the big cities in this section are excellent for poultry farming, as they are so far removed from the great farm region that their bulk of imported eggs are of necessity somewhat stale. California is good chicken country. The Puget Sound country is rather too damp. In the interior western regions the chicken business has not done well, chiefly because the atmosphere is too dry for the methods of artificial incubation attempted.

Section 3 is the great granary of the world. It is also the home of three-fourths of the country's poultry crop. It is a region of corn, cattle and hogs. Such a country will produce poultry in a very inexpensive manner. But it is not the region for special poultry farms. In the northern portion of this tract, we find a heavy housing expense and much winter labor is necessary. It is a region of high priced lands and labor, and low prices for poultry products.

Even the large cities in this region offer little in the way of demand for high grade poultry products. This is because they are so abundantly surrounded with farms that all produce is moderately fresh and plentiful. There are no successful poultry farms in this section west of the Mississippi. It is the natural location of extensive rather than intensive branches of agriculture. The only type of commercial poultry farming that could succeed in any portion of this section would be a large community of producers who could ship their products out regularly in carload lots. Such development could only take place in the southern portion of this region, for the housing expense is too great for the north. At best the distance from market is a disadvantage, for the rate on eggs just about equals the rate on the quantity of grain necessary to produce them. The added time of shipment is something of a

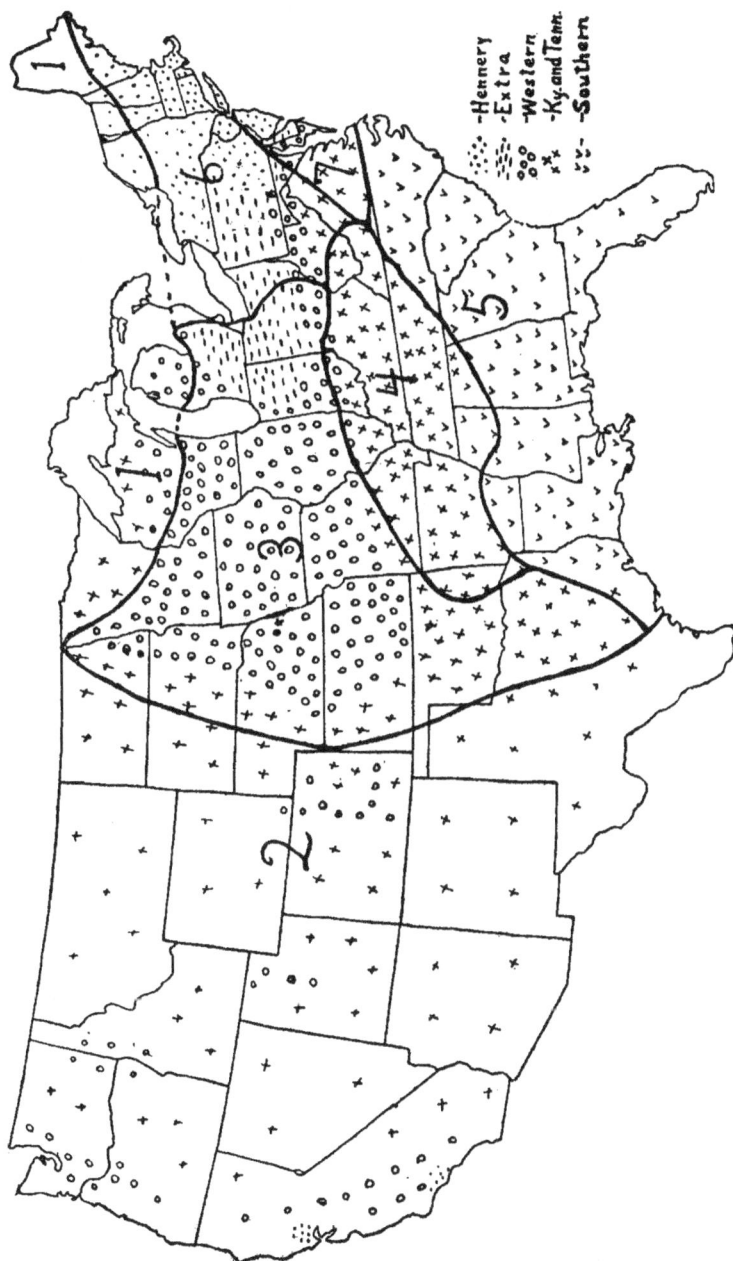

Figure 3. Poultry farming regions.

drawback, though in refrigerator cars this is not serious. After the establishment of poultry communities becomes more common, the Oklahoma and Texas region will become available for this purpose, but they must be established in full swing at the start, for a few isolated poultrymen have no chance at all in this section, for they cannot sell their product to advantage.

Section 4. This region, extending from the Ozarks to Eastern Tennessee, is one of the very best poultry sections. The climate is such that green food is available winter and summer, and the expense of housing and winter labor is reasonable. This section is still in the corn growing region. The question is almost always one of railroad facilities to get the product out. All poultry farms in this section must grow their own grain or buy it of their immediate neighbors. It will not pay to ship grain into this region.

When near shipping facilities, individual poultry farms in Section 4 have a good chance of success, especially east of the Mississippi. This is the most favorable region in the country for the establishment of poultry communities that are to grow their own grain. Such poultry farms will not be expected to confine their attention as exclusively to the business as those in the section where it is profitable to import the grain.[*]

Section 5 is the non-grain growing region of the South. It at present produces little poultry. The climate is all right for the purpose, but the freight rates on grain from the West are high and likewise the freight service and freight rates to the final market are excessive. Under these conditions poultry farming will not pay except in a few localities as in Florida, where there is a high class local market due to the popular resorts. If grain could be profitably grown in this section the same type of poultry farming that prevails in Section 4 would be advisable. Now, grain can be grown in the cotton belt of the South, and many Yankee farmers are making good money doing it. But when grown it is liable to be worth more to feed mules than to feed chickens.

---

[*] Section 4 contains Arkansas, the center of U.S. broiler production.

Section 6 is the "Down East" section of dense population. The land for the most part is rocky, wooded and hilly. The climate and nature of the soil are against the economical production of poultry, but the grain can be profitably fed, and as the markets are the best in the country, commercial poultry farming has gained quite a foothold. If a man is already located in this section and wishes to go into the poultry business I would by all means say, "Go ahead," but I would not advise an outsider looking for a location to come here, for the next section has several advantages.

Section 7 is the best poultry farming district in the United States, either for the individual poultry plant or for the community of poultry growers.* The reasons for this are:

1. The soil is of a sandy nature and excellent land for poultry farming can be had at a low price.
2. The climate is much more favorable than farther north or farther inland.
3. Grain rates from the West are very reasonable.
4. The best market in the country—New York City— is within easy shipping distance.

The type of poultry farming here to he recommended, like that of Section 6, is one in which imported grain is fed. The fertility of this grain, going back on the light soil in the form of manure, is used to grow the green food required by the hens and, in addition, may be used in a rotation system for growing truck crops. It will not pay to grow any quantity of grain. Section 7, because of its advantages over Section 6 in climate and the availability of large tracts of suitable land, is a much better location for the poultry community. Over Section 4, which is the second-best region for this purpose, It has the advantage of nearness to markets. The climatic advantage of Sections 4 and 7 are about on a par. The chief distinction is the matter of growing grain or importing it. If you are to grow your grain, using poultry as a means of

---

* Section 7 contains the Delmarva peninsula, the center of Eastern poultry and egg production. (Hastings' reasoning in this chapter has certainly been borne out by experience!)

marketing it, Section 4 is the best locality. If you are to buy grain, Section 7 is the place.

The boundaries of Section 7 are not arbitrary and should be noted carefully. The line runs from Mattawan, New Jersey, across to the main line of the Pennsylvania and down this to Washington. To the north and west of this, the soils are heavy clays which are wet, cold, slushy and easily befouled. Likewise, the line on the south is distinctly marked by the Norfolk and Western Railway and is a matter of freight rates on grain. Norfolk gets a rate of 16.5¢ from Chicago; a couple of hundred miles south, the rate is about twice as much. Cheaper grain rates would of course extend this belt on down the coast where the climate is even more favorable.

CHICKEN CLIMATE

Climate is a big figure in the cost of poultry production. Every day that water is frozen in winter means increased labor and decreased egg yield. Mild winters mean cheap houses, cheap labor, cheap feed (a large proportion of green food), an earlier chick season, which, together with the mild weather and green feed, mean a large proportion of the egg yield at the season when eggs are high in price.

The American poultry editor wastes a great deal of ink explaining why the Australian egg records of 176 eggs per hen cannot be so; because, in this country, the hens at the Maine station only averaged 126. The Maine Experiment Station lies buried in a snow drift for about five months of the year. The Australian station has a winter climate equal to that of New Orleans. The Australian records do not go below thirty eggs per day per hundred hens at any time during the year. Our New York and New England records run down anywhere from one to ten eggs per day per hundred hens.

Figure 4 will show the effect of the climate upon the distribution of the egg yield throughout the year. The records at New York are from a large number of hens of several different flocks and probably represent a normal distribution of the egg yield for that section. The Kansas and Arkansas lists are taken from the record of small flocks and are not very reliable. The fourth column gives the Australian records with the months transferred on account of being in the southern hemisphere. The last column gives the railroad shipments from a division of the N. C. & St. L. Railroad in Western Tennessee.

Month	Central New York per Hen per Day	Kansas Exp. Station per Hen per Day	Arkansas Exp. Station per Hen per Day	Australian Laying Contest per Hen per Day	Shipments from New Hampshire Egg Farm	Shipment from Western Tennessee
January	.21	.25	.32	.51	26	1509
February	.26	.22	.30	.66	41	1520
March	.43	.60	.62	.67	66	2407
April	.56	.52	.38	.61	83	1775
May	.59	.57	.44	.53	81	1650
June	.50	.46	.42	.45	61	1131
July	.44	.43	.34	.43	58	878
August	.37	.32	.38	.41	54	422
September	.26	.28	.29	.29	24	100
October	.17	.13	.22	.31	3	541
November	.08	.06	.18	.31	2	703
December	.14	.25	.15	.40	11	1150

Figure 4. Egg production in different climates

An equable climate the year round is the best for the chicken business. The California coast is fairly equable in temperature, but its winter rains and summer drought are against it. The Atlantic coast south of New York is fairly good, probably the best the country affords. The most southern portions will be rather too hot in summer, which will result in a small August and September egg yield. Probably the region around Norfolk is, all considered, the best poultry climate the country affords.

SUITABLE SOIL

Soil is important in poultry farming; in fact it is very important, and many failures can be traced to soil mistakes. Rocky and uncultivated lands must not be chosen. To locate on any soil which will not utilize the droppings for the production of green food, is to introduce a loss sufficient to turn success into failure.

The ideal soil for poultry is a soil too sandy to produce ordinary farm crops successfully, and hence an inexpensive soil; but because land too sandy for heavy farming is best for poultry, this does not mean that any cheap soil will do. A heavy wet clay soil worth $150 an acre for dairying is worth nothing for poultry.[*] Pure sand is likewise worthless, and nothing can be more pitiable than to see poultry confined in yards of windswept sand, without a spear of anything green within half a mile.

The soils that are valuable for early truck farming are equally valuable for poultry. Sand with a little loam, or very fine sand, if a few green crops are turned under to provide humus, are ideal poultry soils. The Norfolk Fine Sand and Norfolk Sandy Loam of the U.S. soil survey are types of such soil.

These soils absorb the droppings readily and are never covered with standing water. The winter snows do not stay on them. Crops will keep greener on them in winter than on clay soils three hundred miles farther south.

The disadvantage of such soils is that they lose their fertility by leaching. The same principles that will cause the droppings to disappear from the top of the ground will likewise cause them to be washed down beyond the depths of plant roots. This loss must be guarded against by not going to the extreme in selecting a light

---

[*] Chickens quickly scratch the grass around their houses to pieces, and on a clay soil the bare areas will become quagmires.

soil, and may be largely overcome by schemes of running the poultry right among growing crops, or by quick rotations.

Land sloping to the southward is commonly advised for the purpose of getting the same advantages as are to be had in a sandy soil. In practice the slope of the land cannot be given great prominence, although, other things being equal, one should certainly not disregard this point. In heavy lands it is necessary to raise the floors and grade up around the houses. The quickly drained soil does away with this expense.

Timber on the land is a disadvantage. Poultry farming in the woods has not been made a success. It's the same proposition of the droppings going to waste. I know a man who bought a timbered tract because it was cheap and who scraped up the droppings to sell by the barrel to his neighbor, who used them to fertilize his cabbage patch and in turn sold the poultryman cabbages to feed his hens, at 6¢ a head. Of course, this man failed, as does practically every man who attempts to scrape dropping boards and carry poultry manure around in baskets, instead of using it where it falls.

There is little to be said in favor of uncleared land for the poultry business, but there is something that can be said in favor of the poultry business for uncleared land. A man who buys timbered land for truck farming can get no income whatever the first year, but the poultryman can begin his operations in the woods, clearing the land while he is raising a crop of chickens on it. The coops may be placed in the cleared streak and most of the droppings utilized. In fact, the plan of a streak of timber alongside the houses is not bad for a permanent arrangement—the birds certainly enjoy the shade. But the shade of growing crops is the most profitable kind for poultry.

MARKETING—TRANSPORTATION

The possibilities of working up a local trade of high grade eggs at fancy prices varies greatly with the locality. Large cities and wealthy people are essentials. Other than this the principal distinctions are that regions where a general surplus of eggs are produced offer little chance for a fancy trade. Where the great bulk of eggs are imported fancy trade is more feasible. St. Louis is the smallest western city that supports anything like a fancy trade in eggs and there it is only on a small scale. Minneapolis, Omaha, etc., would not pay 3¢ premium for the best eggs produced, but cities of the same size east of the Appalachians, and especially in New England, will pay a good premium. The Far West or the mountain districts will pay up better than the Mississippi Valley.

The South will pay a little better than the upper Mississippi Valley, but has few cities of sufficient size to make such markets abundant. The Southerner has little regard for quality in produce and the most aristocratic people consume eggs regularly that the wife of a Connecticut factory hand wouldn't have in the house. The egg farmer who expects to sell locally had best not locate south of Washington or west of Pittsburgh, unless he goes to the Pacific Coast.[*]

Where marketing is not done by wagon the subject of railroad transportation is practically identical with the question of marketing. It is the cost in freight service and freight rates that count. The proposition of transportation, especially for the grain-buying poultry farm, catches us coming and going and both must be considered.

A poultry farm in Section 7 will buy one hundred pounds of feed per year per hen and market one-third of a case of eggs. On this basis the grain rate from Chicago or St. Louis and the egg rate to New York must be balanced against each other. Don't take these things for granted. Look them up.

---

[*] This situation has changed considerably since Hastings' time, since there are now gourmets and health-food customers in every community, but in broad outline his description is still true. Marketing is easier in places where people are urban, affluent, or educated. In traditional farm country, as Hastings says, people are used to inexpensive produce and are more reluctant to pay premium prices. This implies a need for better distribution of high-grade products, to get the products to where people will pay top dollar for them.

Jamesburg and Freehold, two New Jersey towns ten miles apart and equidistant and with equal freight rates from New York, might seem to the uninitiated as equally well situated to poultry farming. We will suppose two men bought forty-acre farms of equal quality and equidistant from the railroad stations at these two towns. Suppose, further, they each kept five thousand hens. Jamesburg is on a Philadelphia-New York line of the Pennsylvania and its Chicago grain rate is the same as that of New York, namely: 19½¢ per hundred. Freehold is on a branch line; its rate is 24½¢. In a year the difference amounts to $250. Figured at six percent interest, the land at Jamesburg is worth just about one hundred dollars an acre more than that at Freehold.

Lumber rates or local lumber prices should also be taken into consideration. Whether one plans to ship his product out by express or freight will, of course, be an important consideration in deciding the location.

As a general thing, the individual poultry farmer will, for shipping his product, use express east of Buffalo and north of Norfolk. The poultry community could use freight in these same regions and get as good or better service than by express.

The location in relation to the railroad station is equally important to the freight rate. Besides heavy hauling, frequent trips will be necessary in marketing eggs. These on the larger farms will be daily or at least semi-weekly. On the heavy hauling alone, at 25¢ per ton mile, distance from the railroad will figure up 1¼¢ per hen which, on the basis of the previous illustration, would make a difference of twenty-five dollars per acre for every mile of distance from the station. One of the most successful poultry farms I know is right along the railroad and has an elevator which handles the grain from the cars and later dumps it into the feed wagons without its ever being touched by hand. The labor saving in this counts up rapidly.[*]

---

[*] At Norton Creek Farm, our lives took a turn for the better when we began buying our feed from a mill that delivered directly to our barn.

The poultry community can have its own elevator and the grain can be sold to the farmer to be delivered directly into the hoppers in his field with but a single loading into a wagon.

AVAILABILITY OF WATER

One more point to be considered in location is water.

The labor of watering poultry by carrying water in buckets is tremendous and not to be considered on any up-to-date poultry plant. Watering must be accomplished by some artificial piping system or from spring-fed brooks. The more length of flowing streams on a piece of land, provided the adjacent ground is dry, the more value the property has for poultry. Two spring-fed brooks crossing a forty-acre tract so as to give a half mile of running water, or a full mile of houses, would water five thousand hens without labor. This would mean an annual saving of at least one man's time as against hand watering, or a matter of a thousand dollars or more in the cost of installation of a watering system.

If running water cannot be had the next best thing is to get land with water near the surface which may be tapped with sand points. If one must go deep for water, a large flow is essential so that one power pump may easily supply sufficient water for the plant.

The land should lay in a gentle slope so that water may be run over the entire surface by gravity. Hilly lands are a nuisance in poultrykeeping and raise the expense at every turn.

A FEW STATISTICS

The following table does not bear directly upon the poultryman's choice of a location, but is inserted here because of its general interest in showing the poultry development of the country.

It will be noted that the egg production per hen is very low in the Southern States. This may seem at variance with my previous statements. The poor poultrykeeping of the South is a fault of the industrial conditions, not of the climate. Chickens on the Southern farm simply live around the premises as do rats or English sparrows. No grain is grown; there are no feed lots to run to, no measures are taken to keep down vermin, and no protection is provided from wind and rain. In the North chickens could not exist with such treatment.

The figures given showing the relation between the poultry and total agricultural wealth is the best way that can be found to express statistically the importance of poultrykeeping in relation to the general business of farming. These figures should not be confused with the distribution of the actual volume of poultry products. Iowa, the greatest poultry producing state, shows only a moderate proportion of poultry to all farm wealth, but this is because more agricultural wealth is produced in Iowa than in all the "Down East" states.

State	Eggs Eaten per Capita	% of Farm Wealth Earned by Poultry	No. of Eggs per Hen	Farm Value of a doz. Eggs
Alabama	124	4.9	48	9.7¢
Arizona	80	4.5	60	19.9
Arkansas	235	6.8	58	9.1
California	197	5.4	74	15.8
Colorado	127	5.4	71	15.0
Connecticut	105	11.3	89	19.1
Delaware	231	147	68	13.7
Florida	96	8.2	46	13.1
Georgia	156	4.4	41	10.4
Idaho	213	5.0	67	16.2
Indiana	338	10.0	77	10.5
Iowa	536	7.4	64	10.1
Illinois	215	3.7	62	10.3
Kansas	597	8.2	73	9.9
Kentucky	198	8.3	62	9.8
Louisiana	111	4.0	40	10.0

Figure 5. The development of the poultry industry in the various states, according to the returns of the census of 1900

State	Eggs Eaten per Capita	% of Farm Wealth Earned by Poultry	No. of Eggs per Hen	Farm Value of a doz. Eggs
Maine	233	11.0	100	15.3
Maryland	126	10.4	71	12.6
Massachusetts	56	11.7	96	19.9
Michigan	270	9.7	82	11.2
Minnesota	296	5.8	67	10.5
Mississippi	144	4.7	43	9.9
Missouri	291	11.6	68	9.8
Montana	148	4.3	67	21.0
Nebraska	463	6.1	66	9.9
Nevada	68	3.7	71	20.8
New Hampshire	238	11.5	96	17.3
New Jersey	76	12.0	72	16.2
New Mexico	45	2.7	65	18.7
New York	102	7.1	83	13.9
North Carolina	112	5.7	55	10.2
North Dakota	249	2.6	64	10.5
Ohio	266	9.6	77	11.2
Oklahoma	315	6.4	60	9.3
Oregon	224	6.2	72	15.1
Pennsylvania	112	10.8	75	18.5
Rhode Island	90	19.2	77	20.4
South Carolina	80	4.0	41	10.3
South Dakota	602	5.2	68	10.0
Tennessee	189	8.4	61	9.8
Texas	228	4.8	52	8.0
Utah	146	6.1	76	12.5
Vermont	219	7.5	94	15.3
Virginia	165	8.9	67	11.1

Figure 5. The development of the poultry industry in the various states, according to the returns of the census of 1900 (Continued)

State	Eggs Eaten per Capita	% of Farm Wealth Earned by Poultry	No. of Eggs per Hen	Farm Value of a doz. Eggs
Washington	171	7.1	74	16.8
West Virginia	216	10.2	74	10.9
Wisconsin	268	7.1	68	10.5
Wyoming	121	2.4	79	17.4
Entire U.S.	**205**	**7.4**	**65**	**11.1**

Figure 5. The development of the poultry industry in the various states, according to the returns of the census of 1900 (Continued)

Chapter 5. The Dollar Hen Farm

As has already been emphasized, the way to get money out of the chicken business is not to put so much in.

Land, however, well suited to the purpose, should not be begrudged, for interest at six percent will afford a very considerable extra investment in land well suited to the business if it in any way cuts down the cost of operation.

THE PLAN OF HOUSING

The houses are the next consideration. On most poultry farms they are the chief items of expense.

I know of a poultry farm near New York City where the houses cost $12.00 per hen.[*] The owner built this farm with a view of making money. People also buy stock in Nevada gold mines with a view of making money.

I know another poultry farm owned by a man named Tillinghast at Vernon, Connecticut, where the houses cost 30¢ per hen.[†] Mr. Tillinghast gets more eggs per hen than the New York man. Incidentally, he is sending his son to Yale, and he has no other visible means of support except his chicken farm.

For the region of light soils and the localities which I have recommended for poultry farming, the following style of poultry house should be used:

No floors, single-boarded walls, a roof of matched cypress lumber or of cheap pine covered with tarred paper. This house is to have no windows and no door. The roosts are in the back end; the front end is open or partly open; feed hoppers and nests are in the front end. The feed hoppers may be made in the walls, made loose to set in the house, or made to shed water and placed out-

---

[*] Roughly $240 per hen in 2003 dollars.
[†] Roughly $6 per hen in 2003 dollars.

side the house. All watering is to he done outside the houses; like-wise any feeding beyond that done in hoppers.

The exact style of the house I leave to the reader's own plan. Were I recommending complex houses costing several dollars per hen, this certainly would be leaving the reader in the dark woods. With houses of the kind described it is hard to go far amiss. The simplest form is a double-pitched roof, the ridge-pole standing about seven feet high, and the walls about four. The house is made eight by sixteen, and one end—not the side—left open. For the house that man is to enter, this form cannot be improved upon

The only other points are to construct it on a couple of 4x4 runners so that it can be dragged about by a team of horses. Cypress or other decay-proof wood should be used for these mud-sills. The framing should be light and as little of it used as is consistent with firmness. If the whole house costs more than twenty-five dollars, there is something wrong in its planning. This house should accommodate seventy-five or eighty hens.[*]

For smaller operations, especially for horseless, or intensive farming, a low, light house may be used, which the attendant never enters. A portion of the roof lifts up to fill feed-hoppers, gather eggs or spray. These small houses may be made light enough to be moved short distances by a pry-pole, the team being required only when they are moved to a new field.

Not one particle of poultry manure is to be removed from either style of house. Instead, the houses are removed from the manure, which is then scattered on the neighboring ground with a fork, or, if desired to be used on a field in which poultry may not run, it may be loaded upon a wagon together with some of the underlying soil.[†]

---

[*] One should be able to build such a house today for under $300, using plywood walls and corrugated steel roofing.

[†] A rear scraper blade on a tractor spreads the manure over the surrounding area surprisingly well, hardly disturbing the turf.

There have been books and books written on poultry houses, but what I have just given is sufficient poultry-house knowledge for the Dollar Hen man. If he hasn't enough intelligence to put this into practice, he has no business in the hen business. Additional book-knowledge of hen-houses is useless; it may be harmful.

If you are sure that you are fool-proof, you may get Dr. Feather or Reverend Earlobe's "Book of Poultry House Plans." It will be a good textbook for the children's drawing lessons.

THE FEEDING SYSTEM

Oyster shells, beef scraps, corn, and one other kind of grain, together with an abundance of pasturage or green feed, is the sum and substance of feeding hens on the Dollar Hen Farm.

The dry feeds are placed in hoppers. They are built to protect the feeds from the weather. The neck must be sufficiently large to prevent clogging, and the hopper so protected by slats in front that the hen cannot toss the feed out by a side jerk of her head. These hoppers may be built any size desired. The grain compartments should, of course, be made larger than the others. Weekly filling is good, but where a team is not owned, it would be better to have the hoppers larger so that feed purchased, say, once a month, could he delivered directly into the hoppers.[*]

---

[*] I recommend replacing the beef scrap with a high-protein balanced poultry ration with at least 18% protein, such as a 20% protein layer pellet or a 19% all-purpose poultry feed. Beef scrap these days is of much lower quality than it was in 1909, and carefully formulated poultry rations are the only plausible alternative. Keep the corn, 2nd grain, and oyster shell hoppers.

WATER SYSTEMS

The best water system is a spring-fed brook.

The man proposing to establish an individual poultry plant, and who after reading this book goes and buys a tract of land where an artificial water system is necessary, would catch Mississippi drift-wood on shares. But there are plenty of such people in the world. A man once stood all day on London Bridge hawking gold sovereigns at a shilling apiece and did not make a sale.

Next to natural streams are man-made streams. This is the logical watering method of the community of poultry farmers. These artificial streams are to be made by conducting the water of natural streams back of the land to be watered, as in irrigation. It is the problem of irrigation over again. Indeed, where truck farming is combined with poultry-growing, fowl watering should be combined with irrigation.

It may be necessary to dam the stream to get head, sufficient supply or both. In sandy soils, ditches leak, and board flumes must be substituted. The larger ones are made of the boards at right angles and tapered so that one end of one trough rests in the upper end of the next lower section. The smaller, or lateral troughs may be made V-shaped.

The cost of the smaller sized flume is 3¢ a foot. Iron pipe costs 12¢ a foot.[*]

The greater the slope of the ground the smaller may be the troughs, but on ground where the slopes are great, more expense will be necessary in stilting the flumes to maintain the level, and the harder it will be to find a large section that can be brought under the ditch.

---

[*] Modern plastic irrigation tubing is much cheaper than the pipe available in Hastings' day, and will be cheaper than flumes. I have watered hens both from brooks and from piped water systems, and there is no question that the brooks save a lot of labor compared to pipes, which break down at irregular intervals.

Fluming water for poultry is, like irrigation, a community project. The greatest dominating people of history have their origin in arid countries. It was cooperate or starve, and they learned cooperation and conquered the earth. If a man interferes with the flume, or takes more than his share of the water, put him out. We are in the hen, not the hog business.

Community water systems, where water must be pumped and piped in iron pipe, is of course a more expensive undertaking. It will only pay where water is too deep for individuals to drive sand points on their own property. There is certainly little reason to consider an expensive method when there are abundant localities where simple plans may be used.

On sand lands, with water near the surface, each farmer may drive sand points and pump his water by hand. In this case running water is not possible, but the pipes or flumes may be arranged so that fresh pumping flushes all the drinking places and also leaves them full of standing water. The simplest way to arrange this will be by wooden surface troughs as used in the fluming scheme. The only difference is that an occasional section is made deeper so that it will retain water.

A more permanent arrangement may be made by using a line of three-fourths inch pipe. At each watering place the pipe is brought to the surface so that the water flows into a galvanized pan with sloping sides. This pan has an overflow through a short section of smaller tubing soldered to the side of the pan. The pipe line is parallel with the fence line, the pans supply both fields. By this arrangement the entire plant may be watered in a few minutes. The overflow tubes are on one side. Using these tubes as a pivot the pans may be swung out from under the fence with the foot and cleaned with an old broom. Where the ground water is deep a wind mill and storage tank would be desirable.

OUTDOOR ACCOMMODATIONS

The hen house is a place for roosting, laying and a protection for the feed. The hen is to live out doors.

On the most successful New England poultry farms, warm houses for hens have been given up. Hens fare better out of doors in Virginia than they do in New England, but make more profit out of doors anywhere than they will shut up in houses. If your climate will not permit your hen to live outdoors get out of the climate or get out of the hen business.

There is, however, a vast difference in the kind of out-of-doors. The running stream with its fringe of trees, brush and rank growing grass, forms daylight quarters for the hen par excellence. Rank growing crops, fodder piled against the fences, a board fence on the north side of the lot, or little sheds made by propping a platform against a stake, will all help. A place out of the wind for the hens to dust and sun and be sociable is what is wanted, and what must be provided, preferably by Nature, if not by Nature then by the poultryman.

The hens are to be kept as much as possible out of the houses, in sheltered places among the crops or brush. They should not herd together in a few places but should he separated in little clumps well scattered over the land. These hiding places for the hens must, of course, not be too secluded or eggs will be lost.

EQUIPMENT FOR CHICK REARING

Just as the long houses for hens have been weighed and found wanting, so larger brooder houses, with one exception, have never been established on what may be called a successful basis. By establishment on a successful basis, I mean established so that they could be used by large numbers of people in rearing market chickens. There are plenty of large brooder houses in use, just as there are plenty of yarded poultry plants, but many intelligent, industrious people have tried both systems only to find that the cost of production exceeds the selling price. This makes us prone to believe that some of those who claim to be succeeding may differ from the crowd in that they had more capital to begin with and hence last longer.

The one exception I make to this is that of the South Shore Roaster District of Massachusetts. Here steam-pipe brooder houses are used quite extensively. The logical reason that pipe brooder houses have found use in the winter chicken business and not in rearing pullets is that of season and profits. When chicks are to be hatched in the dead of winter the steam-heated brooder house is a necessity. In this limited use it is all right, where the profits per chick are great enough to stand the expense and losses.

For the rearing of the great bulk of spring chicks the methods that have proven profitable are as follows:

First: Rearing with hens as practiced at Little Compton. For suggestions on this see the chapter entitled "Poultry on the General Farm."

Second: Rearing with lamp brooders.[*] Many large book-built poultry plants have been equipped with steam, or, more properly, hot water heated brooder houses, only to have a practical manager see that they did not work, tear out the piping and fill the house with rows of common lamp brooders. The advantage claimed for the lamp brooder is that they can be regulated separately for each flock. As a matter of fact, the same regulation for each flock of chicks could be secured with a proper type of hot water heaters and one of the most practical poultry farms in the country is now installing such a system.

A brooder system where hot air under the pressure of a blower or centrifugal fan would seem ideal. So far the efforts made along these lines have been clumsy and unnecessarily expensive. If the continuous house is ever made practical, I believe it will be along this line, but at present I advise sticking to the methods that are known to be successful.

Individual lamp brooders in colony houses are perhaps the most generally successful means of rearing chicks on northern poultry farms. They are troublesome and somewhat expensive, but with properly hatched chickens are more successful than hen rearing. In buying such a brooder the chief points to be observed

---

[*] Brooders heated with kerosene lamps lasted until the Twenties, when coal, gas, and electric brooders were introduced. Probably no one misses kerosene or coal brooders, which were labor-intensive and unreliable.

are a good lamp, a heating device giving off the heat from a central drum, and an arrangement which facilitates easy cleaning. The brooder should be large, having not less than nine square feet of floor space. The work demanded of a brooder is not as exacting as with an incubator. The heat and circulation of air may vary a little without harm, but they must not fail altogether. The greatest trouble with brooders in operation is the uncertainty of the lamp. The brooder-lamp should have sufficient oil capacity and a large wick. Brooder-lamps are often exposed to the wind, and, if cheaply constructed or poorly enclosed, the result will be a chilled brood of chicks, or perhaps a fire.

The chief thing sought in the internal arrangements of a brooder is a provision to keep the chicks from piling up and smothering each other as they crowd toward the source of heat. This can be accomplished by having the warmest part of the brooder in the center rather than at the side or corner. If the heat comes from above and a considerable portion of the brooder be heated to the same temperature, no crowding will take place.

The temperature given for running brooders vary with the machine and the position of the thermometer. The one reliable guide for temperature is the action of the chicks. If they are cold they will crowd toward the source of heat; if too warm they will wander uneasily about: but if the temperature is right, each chick will sleep stretched out on the floor. The cold chicken does not sleep at all, but puts in its time fighting its way toward the source of heat. In an improperly constructed or improperly run brooder the chicks go through a varying process of chilling, sweating and struggling when they should be sleeping, and the result is puny chicks that dwindle and die.

The arrangement of the brooder for the sleeping accommodations of the chicks is important, but this is not the only thing to be considered in a brooder. The brooder used in the early season, and especially the outdoor brooder, must have ample space provided for the daytime accommodation of the chick. In the colony house brooder such space will, of course, be the floor of the house.[*]

When operating on a large scale it will not pay to buy complete brooders. The lamps and hovers can be purchased separately and installed in colony houses which do both for breeders and later for houses for growing young stock. The universal hover sold by the Prairie State Incubator people is about as perfect a lamp hover as can be made.

The cold brooder, or Philo box, as it has been popularly called, is the chief item in a system of poultrykeeping that has been widely advertised. The principle of the Philo box is that of holding the air warmed by the chick down close to them by a sagging piece of cloth. The cloth checks most of the radiating heat, but is not so tight as to smother the chick. This limits the space of air to be warmed by the chicks to such a degree that the body warmth is used to the greatest advantage. That chickens can be raised in these fire-less brooders is not in question, for that has been abundantly proven, but most poultryman believe that it will pay better, especially in the North, to give the little fellows a few weeks' warmth.

Curtis Bros. at Ransomville, N.Y., who raise some twenty thousand chicks per year, have adopted the following system: The chicks are kept under hovers heated by hot water pipes for one week, or until they learn to hover. Then they are put in Philo boxes for a week in the same building but away from the pipes. The third week the Philo boxes are placed in a large, unheated room. After that they go to a large Philo box in a colony house.[†]

---

[*] Brooder houses designed for kerosene lamps tended to be very tiny and crowded. Later practice was to give ½ square foot per chick to four weeks and I square foot afterwards.

[†] We are experimenting with a broadly similar system on our farm, using an insulated but unheated brooder for a few weeks after the chicks are put into colony houses on pasture.

To make a Philo house of the Curtis pattern, take a box 5 in. deep and 18 in. to 24 in. square. Cut a hole in one end for a chick door, run a strip of screen around the inside of the box to round the corners. Now take a second similar box. Tack a piece of cloth rather loosely across its open face. Bore a few augur holes in the sides of either box. Invert box No. 2 upon box No. 1. This we will call a Curtis box. It costs about 15¢ and should accommodate fifty to seventy-five chicks.

A universal hover in a colony coop or colony house, for which a Curtis box is substituted, as early in the game as the weather permits, is the method I advise for rearing young chicks. The lamp problem we still have with us, but it is one that cannot be easily solved. Large vessels or tanks of water which are regularly warmed by injection of steam from a movable boiler, offers a possible way out of the difficulty. On a plant large enough to keep one man continually at this work, this plan might be an improvement over filling lamps, but for the smaller plant it is lamps, or go south.

Rearing young chicks is the hardest part of the poultry business. There is a lot of work about it that cannot be gotten rid of. Little chicks must be kept comfortable and their water and feed for the first few days must needs be given largely by hand. They are to be early led to drink from the regular water vessels and eat from the hoppers, but this takes time and patience.

The feeding of chicks I will discuss in the chapter on "Poultry on the General Farm," and as the same methods apply in both cases, I will refer the reader to that section.

After chicks get three or four weeks old their care is the simplest part of the poultry farm work and consists chiefly of filling feed hoppers and protecting them from vermin and thieves.

Board-floor colony houses are used as a protection against rats and this danger necessitates the protection of the opening with netting and the closing of the doors at night.

Cockerels must be gotten out of the flocks and sold at an early age. Those that are to be kept for sale or use as breeders should be early separated from the pullets.

Coops for growing chickens, especially Leghorns, cannot be put among trees, as the birds will learn to roost in the trees, causing no end of trouble to get them broken of the habit.

All pullets save a few culls should be saved for laying. They are to be kept two years. They should lay sixty-five to seventy percent as many eggs the second year as the first. They are sold the third summer to make room for the growing stock.

TWENTY-FIVE ACRE POULTRY FARMS

This section will be devoted to a general discussion of the type of poultry farms best suited to Section 4 and the southerly portions of Section 7 as discussed in the previous chapter.

We will discuss this type of farm with this assumption: That they are to be developed in large numbers by cooperative or corporate effort. This does not infer that they cannot be developed by individual effort, and nine-tenths of the operations will remain the same in the latter case.

Suppose a large tract of land adjacent to railroad facilities has been found. The land in the original survey should be divided into long, relatively narrow strips, lying at right angles to the slope of the land. The farmstead should occupy the highest end of the strip. For a twenty-five acre, one-man poultry farm these strips should be about forty rods (660 feet) in width. The object of this survey is to permit the water being run by gravity to the entire farm.

The first thing is the farmstead, including such orchard and garden as are desired. This stretches across the entire front end of the place. The remainder of the strip is fenced in with chicken fence. The farm is also divided into two narrow fields by a fence down the center of the strip. This fence, at frequent intervals, has removable panels.

The year's season we will begin late in the fall. All layers are in field No. 1 pasturing on rape, top turnips or other fall crops. In lot No. 2 is growing wheat or rye. As the green feed gets short in the first lot the hens are let into lot No. 2. Sometime in March the houses that have been brought up close to the gaps are drawn through into the wheat field. The feed hoppers are also gradually moved and the hens find themselves confined in lot No. 2 without any serious disturbance.

Lot No. 1 is broken up as soon as weather permits and planted in oats, corn, sorghum and perhaps a few sunflowers. The oats form a little strip near the coops and watering places and the sorghum is on the far side.

As soon as corn planting is over the farmer begins to receive his chicks from the hatchery. The brooder coops are now placed in the corn field. The object of the corn is not green food but shade and a grain crop.

The chicks are summered in the corn field and the hens in the wheat or rye. Whether the latter will head up will depend upon the number of the flock. It will be best to work the houses across to the far side and let that portion near the middle fence head up. As the old grain gets too tough for green food strips of ground should be broken up and sown in oats. The grain that matures will not be cut but the hens will be allowed to thresh it out. The straw may be cut with mower or scythe for use as nesting material.

Sometime in June or early in July a little rape, vetch or cow-peas is drilled in between the rows of corn as on the far side from the chicken coops. During July or about the first of August, after all cockerels have been sold, the gates are opened and the pullets are allowed to associate with the hens. After this acquaintance ripens into friendship the hen houses are worked back into the pullet lots. Surplus hens are sold off or new houses inserted as the case may be until there is room for the pullets in the houses. Each brooder coop is worked up alongside a house and after most of the pullets have taken to the houses the coops are removed. The vacant lot is now broken up and sown in a mixture of fall green crops.

The flock is kept in the corn field until the corn is ripe. The sorghum and sunflowers are knocked down where they stand and are threshed by the hens. As soon as the corn crop is ripe the houses are run back and the corn cut up or husked and the wheat planted in the corn field.

The next year the lots are transposed, the young stock being grown in the lot that had the hens the previous year.

If the ground is inclined to be at all damp when the fields are broken up the plowing is done in narrow lands so as to form a succession of ridges, on which are placed the coops or houses. The directions of these ridges will be determined by the lay of the land—the object being neither to dam up water or to encourage washing. The location of the ridges are alternated by seasons, so that the droppings from the houses are well distributed through-out the soil.

This system, with the particular crops found to do best in the locality, give us an ideal method of poultry husbandry. We have kept hens and young stock supplied with green food the year round; we have utilized every particle of manure without one bit of labor. We have a rotation of crops. We have the benefits to the ground of several green crops turned under. We have raised one grain crop per year on most of the ground. We have no labor in feeding and watering except the keeping of the grain, beef and grit hoppers filled, and the water system in order.

The number of fowls that may be kept per acre will be determined by the richness of the soil. The chief object of the entire scheme is to provide abundant green pasture at all times and to allow the production of a reasonable amount of grain. With one hundred hens per acre on the entire tract, and with houses containing eighty hens each, it will be necessary to set the houses ninety-five feet apart. This will give the flock a tract of 95 by 830 feet in which to pasture.

The above estimate, with a little land allowed for house, gar-den, orchard and a little cow and team pasture, will permit the keeping of two thousand hens on a twenty-five acre farm. In regions where grain is to be raised most farmers would want more land. They may also wish to own a few extra cows, hogs, etc., or to alternate the entire poultry operations with some crop that will, on such highly fertilized land, give a good cash profit. Forty acres is a good size for such uses.

The cost of land when purchased in large tracts in Virginia is very small, but the cost of clearing is often much more than that of the land. Twenty-five to fifty dollars an acre should secure such a tract of land and put it in shape for poultry farming.

The cost of the farm home, etc., will, of course vary altogether with the taste of the occupant. If they are constructed by a central company, from five hundred to a thousand dollars should cover the amount.

The cost of poultry buildings and equipment used on the farm will depend largely on the efficiency of the labor of construction. If constructed in large numbers by a central company, the cost would be reduced, but the company would expect to make a profit on their work.

A plot laid out for two thousand hens will require in material: 250 rods (4,125 feet) of fence with 6-ft. netting which should cost about 50¢ a rod. My estimate of this fence put up would be $150. If the neighboring field contained no other poultry, a portion of this fence might be done away with, although its protection against dogs and strangers may be worth while. Of course, if poultry fields of different owners lay adjoining, the fence must be used, but the cost will be reduced one-half.[*]

The next most expensive piece of equipment will be a line of about eighty rods (1,320 feet) of ¾ in. gas pipe and about fifty elbows and twenty-five galvanized iron pans. The cost of installation will depend largely on how deep it is necessary to go to get below the frost line. One hundred and seventy-five dollars should cover cost of material and by the use of a plow the line ought to be put in for twenty-five dollars.

The source of water, and the cost of getting a head, will necessarily vary with the location. The installation of a windmill and

---

[*] To hold in poultry and keep out predators, I use two strands of aluminum fence wire, one 5" off the ground and the other 10". This has greatly reduced losses to dogs, coyotes, and raccoons. Such a fence can be stepped over, so one does not have to deal with gates except to get a truck or tractor in. Areas with coyotes that are wise about electric fences will require additional strands of wire higher off the ground. Such a fence is easy to set up and is very inexpensive.

tank to hold a supply for several days, or of a small gasoline engine, would cost in the neighborhood of one hundred dollars, but it is a luxury that may be dispensed with if the well is not too deep.

The houses for the hens, of which there are twenty-five, are constructed in accordance with some of the plans previously discussed. The cost should be about 25¢ per hen.

At least twice as many brooder coops will be needed as there are hen houses, but of the lamps and hovers not over twenty-five will be required, as the chicks soon outgrow the need of this aid.

This makes a list of equipment required for the keeping of two thousand layers and their replenishing;

25 acres of farm land, at $50 per acre	$1250.00
250 rods of fence	150.00
One farmstead	1000.00
One team, plow and farm implements	800.00
One watering system	800.00
25 hen houses, at $20	500.00
50 colony coops, at $2.50	150.00
25 lamps and hovers, at $5	125.00
Total	**$3775.00**

This is a good, liberal capitalization. The business can be started with much less. Figuring interest at 6%, we have $225.00 per year.

The upkeep of the plant will be about 15% of the capital, not counting land. This equals $375, which, added to interest, gives an annual overhead expense of $600, which is our first item to be set against gross receipts.

The cost of operation will involve cost of chicks at hatchery, purchased feed, seed for ground, and feed for team.

The price of chicks at the Petaluma hatcheries is from 6¢-8¢ each. We expect to raise enough pullets to make up for the accidental losses, and to renew bulk of the flock each year. The num-

ber required will, of course, depend upon the loss. This loss will
be much less when the chicks are obtained from a modern moisture controlled hatchery than from the box type incubator. I think
a 33% loss is a liberal estimate, but as I am treading on unproven
ground. I will make that loss 40%, which is on a par with old-style
methods. To replace 1,000 hens, this will require 3,500 chicks at a
cost of about two hundred and fifty dollars.

Green pasturage throughout the year will materially cut down
the cost of feed. The corn consumed out of the hoppers will be
about one bushel per hen. The beef scrap will also be less than
with yarded fowls, perhaps 25¢ per hen. Now, of the corn we will
raise on the land, at least ten acres. This should yield us five hundred bushels. This leaves fifteen hundred bushels of corn to be
purchased. At the present high rates, this will cost $1,000 which,
added to beef scrap cost, makes an outside feed cost of $1,500.
The seed cost of rye, rape, cow-peas, etc., will amount to about
$50 per annum. For expense of production we have:

Interest and upkeep of plant	$600.00
Chicks	250.00
Purchased corn	1000.00
Beef scrap and grit	500.00
Seed	50.00
Team feed	100.00
Total	**$2,500.00**

This figures out the cost of production at a little more than a
dollar per hen. The income from the place should be about as follows: Eleven hundred cockerels sold as squab broilers at 40¢ each,
$440.00; four hundred and seventy-five old hens at 30¢, $140.00.

The receipts from egg yield are, of course, impossible of very
accurate calculation, for it is here that the personal element that
determines success or failure enters. The Arkansas per-hen-day
figures (see last chapter), multiplied by the average quotation for

Extras[*] in the New York market, will be as fair as any, and certainly cannot be considered a high estimate, as it is only 113 eggs per hen per year.

Month	Eggs per hen day	Price per doz. Extras in New York	Income for Month from 2000 Layers
January	.32	$ .30	$494.00
February	.30	.29	404.00
March	.62	.22	100.00
April	.38	.19	350.00
May	.44	.19	429.00
June	.42	.18	371.00
July	.34	.21	367.00
August	.38	.22	429.00
September	.21	.25	202.00
October	.22	.28	310.00
November	.18	.33	267.00
December	.16	.32	246.00
Total			$4,641.00

The total income as figured will be $5,221. From this subtract the cost of production, and we have still nearly $3,000, which is to be combined item of wages and profit. We have entered no labor bill because this is to be a one-man farm, and with the assistance of the public hatchery and cooperative marketing association, which will send a wagon right to a man's door to gather the eggs, it is entirely feasible for one man to attend to two thousand hens. In the rush spring season other members of the family will have to turn out and help, or a man may be hired to attend the plowing and rougher work.

This is a good handsome income, and yet it is only about one dollar per hen, which has always been the estimated profit of suc-

---

* Grade A eggs.

cessful poultrykeeping. As a matter of fact, this profit is seldom reached under the old system of poultrykeeping, not because the above gross income cannot be reached, but because the expenses are greater. Under the present methods, with the exception of the rearing of the young chicks, one man can easily take care of three thousand hens. Indeed, practically the only work in their care is cultivating the ground and hauling around and dumping into hoppers about two loads of feed per week

But young chicks must be reared, and this is more laborious. For this reason I advise going into some other industry on a part of the land which will not require attention in the young chick season. One of the best things for this purpose is the cultivation of cane fruits as blackberries, raspberries and dewberries. The work of caring for these can be made to fall wholly outside the young chick season. Peaches and grapes for a slower profit can be added, but spraying and cultivation these is more liable to take spring labor. All these fruits have the advantage of doing well in the same kind of soil recommended for chickens. Young chickens may be grown around such berry crops and removed to permanent quarters before the berries ripen. Strawberries would be a very poor crop because their labor falls in the chick season.

Another plan, and perhaps a better one, is to have about three fields, and rotate in such a manner that a marketable crop may be always kept growing in the third field. Any crop may be selected, the chief labor of which falls between July and the following March. Late cabbage and potatoes, or celery, will do if the ground is suitable for these crops. Kale and spinach are staple fall crops. Fall lettuce could also be grown. If the market is glutted on such crops, they can be fed to the chickens. Whenever a field is vacant, have some crop growing on it, if only for purposes of green manuring. Never let sandy ground lie fallow.

A modification of the above plans, suited to heavier ground, is to seed down the entire farm to grass. It is then divided into three fields and provided with three sets of colony houses. Coops are entirely dispensed with, and cheap indoor brooders are used in the permanent houses. The pullets stay in the same house in the same field until the moulting season of the third year, or until they are two and a half years old. One field will always be vacant during the fall and winter season which time may be used for fresh seeding.

The difficulty of maintaining a sod will necessitate somewhat heavier soil than by the previous plan. The houses should be moved around occasionally, as the grass kills out in the locality. This plan is a lazy man's way, taking the least labor of any method of poultrykeeping known. It is adapted to the cheaper ground in the region farthest from market.[*] On the Atlantic seaboard, the more enterprising man will use the third field for rotation, and sell some of the fertility of the western grain in the form of a truck crop.

FIVE ACRE POULTRY FARM

Can a living for a family be made from a five acre poultry farm? Yes; by individual effort, where the marketing opportunities are good; or by corporate or cooperate effort, any place where the fundamental conditions are right.

This type of poultry farm is well suited for development near our large cities, where the cry of "back to the land" has filled with new hope the discouraged dweller in flat and tenement. No greater chance for humanitarian work, and at the same time no greater business opportunity, is open today than that of the promotion of colonies of small poultry and truck farms where the parent colony not only sells the land, but helps the settler to establish himself in the business and to successfully market the product. The natural location for such projects is in the sandy soils of New Jersey, Delaware, Maryland and Virginia.

---

[*] I use grass pasture. All my hens are kept on permanent pasture, which is never plowed. My farm is on a clay soil and keeping it in grass prevents problems with mud. Permanent pasture fertilized with chicken manure grows plenty of forage for our chickens, sheep, and goats. The chief disadvantage of this plan is that permanent pasture provides no shade. Chickens like shade, and sources of shade also provide protection from hawks.

We have already discussed the twenty-five acre farm, representing the largest probable unit for such an enterprise. We will now discuss the five acre farm which represents the smallest probable unit.

On the five-acre farm a considerable difference of methods will be necessary. In the first place, it is to be a horseless farm. All hauling and plowing must be attended to by the central company, or the same results could be obtained by a team owned in common by a small group, say of six farmers, each of whom is to use the team one day of the week.

A single isolated farmer in a community of farms or market gardeners could hire a team by the day as he needed it. I do not recommend this scheme, however, but would suggest that the single individual get a larger plot of ground, at least ten acres, and a team of his own. In the cooperative community the five-acre teamless farm is entirely feasible.

The tract should be surveyed about twice as long as wide, which, for five acres, makes it 20 by 40 rods, or 330 by 660 feet Measure off a strip one hundred feet back from the road. Fence the remainder of the tract. Now run a partition fence down the center until we have come to within twelve rods of the back side. Here run a cross fence. This gives us three yards of about one and one-half acres each. The gates are arranged so that one passes through the three yards in a single trip.

Where the middle partition fence adjoins the front fence, a well is driven. A water line is run down the partition fence to the rear yard.

The plot around the house is set in permanent crops, such as berries, fruit trees, asparagus, rhubarb, etc. Of the other three yards, at least one is kept in growing marketable crops. Every inch is cultivated, and crops of the leafy nature, as lettuce, cabbage, kale and spinach, are chiefly grown, as they utilize the rich nitrogenous poultry manure to the best advantage, and the waste portions, or worthless crops, are utilized for the poultry. The method of supplying the fowls with green food is entirely by soiling. This means to grow the food in an adjoining lot and throw it over the fence.

The above mentioned crops are all good for the purpose. Rape, which is not grown for human food, is also excellent.[*]

Kale is one of the very best crops for soiling purposes. It is planted in the fall and fed by pulling off the lower leaves during the winter. In the spring the hardened stalks stand at a considerable height and the field may be used for growing young chicks, giving shade, and at the same time producing abundant green feed, without any immediate labor, which means a great saving in the busy season.

A set of panels or netting stretched on light frames is provided. They are of sufficient number to set along the longest side of one of the fields. A strip along the fence, four or five feet wide, can be planted to sunflowers, corn, rape, kale, or other rank growing crop and the panels leaned against the fence to protect the young plants from the hens. In this way the fence rows can be kept provided with the shade of growing crops, which relieves the otherwise serious fault of this plan of poultry farming, in that the hens would be required to live in absolutely barren and sunburned lots, for we propose to keep five or six hundred hens on one and a half acres of ground, and no green things could get a start without protection.

Rotate the houses from field to field as often as the crops allow. Never permit hens to run in one bare field for more than six months at a time. Always keep every inch of ground not in use by the chickens, luxuriant in something green. If you have a crop of vegetables which are about matured, drill rape or crimson clover between the rows: by the time the crop is harvested and the hens are to be moved in, such crops will have made a good growth. The hens will kill it out but it will be a "profitable killing."

By this system of intensive combination of truck farming and poultry farming, we have a combination which for small capital and small lands cannot be beaten. The hens should yield better than a dollar profit per head on this plan; the one and a half acres

---

[*] Soiling is used because 600 hens on 1 ½ acres will quickly destroy all plant life in their yard.

automatically fertilized and intensely cultivated, growing two or three crops a year, should easily double the income.[*]

Twelve hundred dollars a year is a conservative estimate for the net income from such a plant, and the original investment, exclusive of residence, will not be over one thousand dollars.

---

[*] Hastings understands that whenever you have more than about 100 hens per acre, the ground will quickly become denuded. With large amounts of acreage and portable houses, the hens can be moved to greener pastures as quickly as this occurs. When space is restricted, as in the five-acre farm (or with kind of yarding with permanent houses), things quickly devolve into a mud-yard operation, where the yard is so contaminated that the chickens would be better off in confinement. Even with a triple-yard rotation, as described here, the hens will be in barren yards most of the time, but frequent plowing and replanting keep the contamination problem at bay.

Chapter 6. Incubation

The differences in the process of reproduction in birds and mammals is frequently misunderstood. The laying of the bird's egg is not analogous to the birth of young in mammals.

The female, whether bird or beast, forms a true egg which must be fertilized by the male sperm cell before the offspring can develop. In the mammal, if fertilization does not occur, the egg, which is inconspicuous, passes out of the body and is lost. If fertilized, it passes into the womb where the young develops through the embryonic stages, being supplied with nourishment and oxygen directly by the mother.

In the bird, the egg, fertilized or unfertilized, passes out of the body and, being of conspicuous size, is readily observed. The size of the egg is due to the supply of food material which is comparable with that supplied to the mammalian young during its stay in the mother's womb.

Reptiles lay eggs that are left to develop outside of the body of the mother, subject to the vicissitudes of the environment. The young of the bird, being warm-blooded, cannot develop without a more uniform temperature than weather conditions ordinarily supply. This heat is supplied by the instinctive brooding habit of the mother bird.

FERTILITY OF EGGS

In a state of nature the number of eggs laid by wild fowl are only as many as can be covered by the female. These are laid in the spring of the year, and one copulation of the male bird is sufficient to fertilize the entire clutch. Under domestication, the hen lays quite indefinitely, and is served by the male at frequent intervals. The fertilizing power of the male bird extends over a period of about 15 days.

For most of my readers, it will be unnecessary to state that the male has no influence upon the other offspring than those which he actually fertilizes within this period. The belief in the influence of the first male upon the later hatches by another male is simply a superstition.

The domestic chicken is decidedly polygamous. The common rule is one male to 12 or 15 hens. I have had equally good results, however, with one male to 20 hens. In the Little Compton and South Shore districts, one male is used for thirty or even forty hens.

By infertile eggs is meant eggs in which the sperm cell has never united with the ovum. Such eggs may occur in a flock from the absence of the male, from his disinclination or physical inability to serve the hens, from the weakness or lack of vitality in the sperm cells, from his neglect of a particular hen, from lifelessness, or lack of vitality in the ovule, or from chance misses, by which some eggs fail to be reached by the sperm cells.

In practice, lack of sexual inclination in a vigorous looking rooster is very rare indeed. The more likely explanation is that he neglects some hens, or that the eggs are fertilized, but the germs die before incubation begins, or in the early stages of that process. The former trouble may be avoided by having a relay of roosters and shutting each one up part of the time. The latter difficulty will be diminished by setting the egg as fresh as possible, meanwhile storing them in a cool place. The other factors to be considered in getting fertile eggs are so nearly synonymous with the problems of health and vitality in laying stock generally, that to discuss it here would be but a repetition of ideas.

In connection with the discussion of fertile eggs, I want to point out the fact that the whole subject of fertility, as distinct from hatchability, is somewhat meaningless. The facts of the case are, that whatever factors in the care of the stock will get a large percentage fertile eggs will also give hatchable eggs and vice versa. This is to be explained by the fact that most of the "infertile" eggs tested out during incubation are in reality dead germs in which death has occurred before the chick became visible to the naked eye. Such deaths should usually be ascribed to poor parentage, but may be caused by wrong storage or incubation. Likewise, it would not be just to credit all deaths after chicks became visible to wrong incubation, although the most of the blame probably belongs there.

Likewise, with brooder chicks, we must divide the credit of their livability in an arbitrary fashion between parentage, incubation, and care after hatching.

By the hatchability of eggs, we then mean the percentage of eggs set that hatch chicks able to walk and eat. By the livability of chicks, we mean the percentage of chicks hatched that live to the age of four weeks, after which they are subject to no greater death rate than adult chickens. By the livability of eggs, we mean the product of these two factors, i.e., the percentage of chicks at four weeks of age based upon the total number of eggs set.

As before mentioned, the fertility of eggs bears a fairly definite relation to the hatchability, so likewise the hatchability bears a relation to the livability of chicks. When poor hatches occur because of weak germs, as because of faulty incubation, this same injury to the chick's organism is carried over and causes a larger death among the hatched chicks.

Moreover, the relation between the two is not the same with all classes of hatches, but as batches get poorer the mortality among the chicks increases at an accelerating rates. Figure 6 gives a rough approximation of these ratios.

These figures are based on incubator data. Eggs set under hens usually give a hatchability of 50% to 65%, and livability of 70% to 80%. The reason for the greater livability is that the real hatchability of the eggs is 70% to 75%, and is reduced by mechanical breakage. The hatchability of eggs varies with the season. This variation is commonly ascribed to nature, it being stated that springtime is the natural breeding season, and therefore eggs are of greater fertility.

While there may be a little foundation for this idea, the chief cause is to be found in the manner of artificial incubation, as will be discussed in a later section of this chapter.

Percent of Hatchability	Percent of Chick Livability	Percent of Egg Livability
100	100	100
90	95	85
80	88	70
70	84	50
60	72	48
50	55	27
40	40	16
30	24	7
20	10	2
10	2	1

Figure 6. Relationship between livability and hatchability

The following table is given as the seasonable hatchability for northern states. This is based on May hatch of 50%:

January	38	July	40
February	42	August	40
March	47	September	42
April	49	October	43
May	50	November	40
June	46	December	35

Most people have an exaggerated idea of the hen's success as a hatcher. I have a number of records of hen hatching with large numbers of eggs set, and they are all between 55% and 60%. The reasons the hen does not hatch better are as follows:

1. Actual infertile eggs—usually, running about 10% in the best season of the year.
2. Mechanical breakage.

3. Eggs accidentally getting chilled by rolled to one side of the nests, or by the sick, lousy or crazy hens leaving the nests or standing up on the eggs.
4. Eggs getting damp from wet nests, dung or broken eggs; thus causing bacterial infection and decay.

The last three causes are not present in artificial incubation. From my observation they cause a loss of 15% of the eggs that fail to hatch, when hens are managed in large numbers. This would properly credit our hens with hatches running from 70% to 75%, which, for reasons later explained, is not equal to hatches under the best known conditions of artificial incubation.

The assumption that the hen is a perfect hatcher, even barring accidents and the inherited imperfection of the egg, is not, I think, in harmony with our general conception of nature. Not only are eggs under the hens subject to unfavorable weather conditions, but the hen, to satisfy her whims or hunger, frequently remains too long away from the eggs, allowing them to become chilled.

For directions of how to manage setting hens, consult the Chapter on "Poultry on the General Farm,"

THE WISDOM OF THE EGYPTIANS

Up to the present there have been just three types of artificial incubation that have proven successful enough to warrant our attention. These are:
1. The modern wooden-box-kerosene-lamp incubator which is seen at its best development in the United States.
2. The Egyptian incubator of ancient origin, which is a large clay oven holding thousands of eggs and warmed by smouldering fires of straw.
3. The Chinese incubator, much on the principle of the Egyptian hatchery, but run in the room of an ordinary house, heated with charcoal braziers and used only for duck eggs.

I have no accurate information on the results of the Chinese method, and as it is not used for hen eggs, we will confine our attention to the first two processes only.

I do not care to go into detail in discussing makes of box incubators, but I will mention briefly the chief points in the development of our present machines.

The first difficulties were in getting lamps, regulators, etc., that would give a uniform temperature. This now has been worked out to a point where, with any good incubator and an experienced operator, the temperature of the egg chamber is readily kept within the desired range.

These are two principal types of box incubators now in use. In the earliest of these, the eggs were heated by radiation from a tank of hot water. These machines depended for ventilation or, what is much more important, evaporation, upon chance air currents passing in and out of augur holes in the ends or bottom of the machine.

The second, or more modern type, warms the eggs by a current of air which passes around a lamp flue where, being made lighter by the expansion due to heat, the air rises, creating a draft that forces it into the egg chamber. There it is caused to spread by muslin or felt diaphragms so that no perceptible current of air strikes the eggs. This type is the most popular type of small incubator on the market. Its advantage will be more readily seen after the discussion of the principles of incubation.*

Hazy tales of Egyptian incubators have gone the rounds of poultry papers these many years. More recently some accurate accounts from American travelers and European investigators have come to light, and as a result, the average poultry editor is kept busy trying to explain how such wonderful results can be obtained "in opposition to the well-known laws of incubation."

The facts about Egyptian incubators are as follows: They have a capacity of 50,000 to 100,000 eggs, and are built as a single large room, partly underground and made of clay reinforced with straw.

---

* This type of incubator is analogous to the electric still-air incubators that are still sold in large quantities. All the issues discussed in this chapter are directly applicable to still-air incubators, and are largely applicable to electric cabinet incubators with circulating fans.

The walls are two or three feet thick. Inside, the main rooms are little clay domes with two floors.

The hatching season begins the middle of January and lasts three months. A couple of weeks before the hatching begins, the fireproof house is filled with straw which is set afire, thoroughly warming the hatchery. The ashes are then taken out and little fires built in pots are set around the outside of the big room. The little clay rooms with the double floors are now filled with eggs. That is, one is filled at a time, the idea being to have fresh eggs entering and chicks moving out in a regular order, so as not to cause radical changes in the temperature of the hatchery.

No thermometer is used, but the operator has a very highly cultivated sense of temperature, such as is possessed by a cheese maker or dynamite dryer. About the twelfth day the eggs are moved to the upper part of the little interior rooms where they are further removed from the heated floor. The eggs are turned and tested out much as in this country. They are never cooled and the room is full of the fumes and smoke of burning straw. The ventilation provided is incidental.

This is about the whole story save for results. The incubator men pay back three chicks for four eggs, and take their profits by selling the extra chicks that are hatched above the 75%. This statement is in itself so astonishing and yet convincing, that to add that the hatch runs between 85% and 90% of all eggs set, and that the incubators of the Nile Delta hatch about 75,000,000 chicks a year seems almost superfluous. As for the explanation of the results of the Egyptian incubators compared with the American kerosene lamp type, I think it can best be brought about by a consideration in detail of the scientific principles of incubators.

PRINCIPLES OF INCUBATION

To keep animal life, once started, alive and growing, we need:
1. A suitable surrounding temperature.
2. A fairly constant proportion of water in the body substance.
3. Oxygen.
4. Food.

Now, a fertile egg is a living young animal and as such its needs should be considered. We may at once dispose of the food problem of the unhatched chick by saying that the food is the contents of the egg at the time of laying, and as far as incubation is concerned, is beyond our control.

In consideration of external temperature in its relation to life, we should note:

1. The optimum temperature.
2. The range of temperature consistent with general good health.
3. The range outside which death occurs.

Just to show the principle at stake, and without looking up authorities, I will state these temperatures for a number of animals. Of course you can dispute the accuracy of these figures, but they will serve to illustrate our purpose:

Species	External Optimum Point	External Healthful Range	External Survivable Range	Internal Optimum Point	Internal Survivable Range
Man	70	50 to 100	50 to 140	98	90 to 106
Dog	60	20 to 140	20 to 140	101	96 to 110
Monkey	90	30 to 140	30 to 140	101	95 to 108
Horse	80	20 to 120	20 to 120	99	95 to 105
Fowl	80	20 to 140	20 to 140	107	100 to 115
Newly hatched chick	90	70 to 100	40 to 120	108	100 to 115
Fertile egg at start of incubation	103	32 to 110	31 to 125	103	31 to 125
Egg incubated three days	103	98 to 105	80 to 118	108	95 to 118
Egg incubated eighteen days	102	75 to 105	50 to 118	106	98 to 116

This table shows, among other things, that we are considering in the chick not a new proposition to which the laws of general animal life do not apply, but merely a young animal during the process of growth to a point where its internal mechanism for heat control has power to maintain the body temperature through a greater range of external temperature change.

In the cooling process that occurs after laying the living cells of the egg become dormant, and like a hibernating animal, the actual internal temperature can be subjected to a much greater range than when the animal is active. After incubation begins and cell activity returns, and especially after blood forms and circulation commences, the temperature of the chick becomes subject to about the same internal range as with other warm-blooded animals.

In the case of fully formed animals, the internal temperature is regulated by a double process. If the external temperature is lowered, more food substance is combined with oxygen to keep up the warmth of the body, while if the external temperature be raised, the body temperature is kept low by the cooling effects of evaporation. This occurs in mammals chiefly by sweating. Birds do not sweat, but the same effect is brought about by increased breathing. Now, the chick gradually develops the heat producing function during incubation, until towards the close of the period it can take care of itself fairly well in case of lowered external temperature. The power to cool the body by breathing is not, however, granted to the unhatched chick, and for this reason the incubating egg cannot stand excess of heat as well as lack of it.

The practical points to be remembered from the above are:

1. Before incubation begins, eggs may be subjected to any temperature that will not physically or chemically injure the substance.

2. During the first few days of the hatch, eggs have no appreciable power of heat formation and the external temperature for any considerable period of time can safely vary only within the range of temperature at which the physiological process may be carried on.

3. As the chick develops it needs less careful guarding against cooling, and must still be guarded against over-heating.
4. It should be remembered, however, that eggs are very poor conductors of heat, and if the temperature change is not great several hours of exposure are required to bring the egg to the new temperature.

Temperature is the most readily observed feature about natural incubation and its control was consequently the first and chief effort of the early incubator inventors.

A great deal of experimental work has been done to determine the degree of temperature for eggs during incubation. The temperature of the hen's blood is about 105-107° F. The eggs are not warmed quite to this temperature; the amount by which they fail to reach the temperature of the hens body depending, of course, upon the surrounding temperature. 103° F is the temperature that has been generally agreed upon by incubator manufacturers. Some of these advise running 102° the first week, 103° the second, 104° the third. As a matter of fact it is very difficult to determine the actual temperature of the egg in the box incubator. This is because the source of heat is above the eggs and the air temperature changes rapidly as the thermometer is raised or lowered through the egg chamber. The advice to place the bulb of the thermometer against the live egg is very good, but in practice quite variable results will be found on different eggs and different parts of the machine.

With incubators of the same make, and in all appearances identical, quite marked variation in hatching capacity has been observed in individual machines. Careful experimentation will usually show this to be a matter of the way the thermometer is hung in relation to the heating surfaces and to the eggs. Ovi-thermometers, which consists of a thermometer enclosed in the celluloid imitation of an egg, are now in the market and are perhaps as safe as anything that can be used.

As was indicated in the previous section greater care in temperature of the egg is necessary in the first half of the hatch. The temperature of 102° F as above given is, in the writer's opinion, too low for this portion of the hatch. An actual temperature of

104° at the top of the eggs will, as has been shown by careful experimental work, give better hatches than the lower temperature.

MOISTURE AND EVAPORATION

The subject of the water content of the egg, and its relation to life, is the least understood of poultry problems.

The whole study of the water content of the egg during incubation hangs on the amount of evaporation. Now, the rates of evaporation from any moist object is determined by two factors: vapor pressure and the rate of movement of the air past the object. As incubation is always carried on at the same temperature, the evaporating power of the air is directly proportioned to the difference in the vapor pressure of water at that temperature, and the vapor pressure of the air as it enters the machine. Thus, in order to know the evaporative power of the air, we have only to determine the vapor pressure of the air and to remember that the rate of evaporation is in proportion to this pressure: i.e., when the vapor pressure is high, the evaporation will be slow and the eggs remain too wet, and when the vapor pressure is low the eggs will be excessively dried out.

The reader is probably more familiar with the term "relative humidity" than the term "vapor pressure," but as the actual significance of relative humidity is changed at every change in outside temperature, the use of this term for expressing the evaporating power of the air has led to no end of confusion.

The influence of air currents on evaporation is to increase it directly proportional with the rate of air movement. Thus, 10 cubic feet of air per hour passing through an egg chamber would remove twice as much moisture as would 5 cubic feet.

If the percentage of water in any living body be changed a relatively small amount, serious disturbances of the physiological processes and ultimately death will result. The mature animal can, by drinking, take considerable excess of water without danger, for the surplus will be speedily removed by perspiration and by the secretion from the kidneys. But the percentage of water in the actual tissues of the body can vary only within a narrow range of not more than three or four percent. The chick in the shell is not provided with means of increasing its water content by drinking or diminishing it by excretion, but the fresh egg is provided with more moisture than the hatched chick will require, and the surplus is gradually lost by evaporation. This places the water content of the chick's body at the mercies of the evaporating power of the air that surrounds the egg during incubation.

To assume that these risks of uncertain rates of evaporation is desirable, is as absurd as to assume that the risks of rainfall are desirable for plant life, As the plants of a certain climate have become adapted to the amount of soil moisture which the climate is likely to provide, so the egg has by natural selection been formed with about as much excess of water as will be lost in an average season under the natural conditions of incubation. Plant life suffers in drought or flood, and likewise bird life suffers in seasons of abnormal evaporative conditions. This view is substantiated by the fact that the eggs of water fowl which are in nature incubated in damper places, have a lower water content than the eggs of land birds.

The percent of water contained in the contents of fresh eggs is about 74%, or about 65.5% based on the weight, shell included. Unfortunately no investigations have been made concerning the percent of water present in the newly hatched chick.

Upon the subject of the loss of water for the whole period of incubation, valuable data has been collected at the Utah, Oregon and Ontario Experiment Stations.

In these tests we find that as a rule the evaporation of eggs under hens is less than in incubators. With both hens and incubators, the rate of evaporation is greatest at the Utah Station, which one would naturally expect from the climate. The eggs under hens at the Ontario Station averaged about 12% loss in weight, and those at the Utah Station about 15%. At both stations, incubators without moisture ran several percent higher evaporation than eggs under hens. The conclusions at all stations were that the addition of moisture to incubators was a material aid to good batches of livable chicks.

At Ontario the avenge evaporation ran as low as 1%. At Utah it reached as high as 24%. Now as the entire loss of weight is loss of water, the solid contents remaining the same, and as the original percent of water contained in the egg (shell included) is only 65.5%, the chicks of the two lots with the same amount of solid substance would contain water in the proportion of 58.5 to 41.5. Based on the weight of the chick, this would make a difference of water content of over 25%.

That human beings or other animals could not exist with such differences in the chemical composition of the body, is at once apparent. In fact I do not believe that the chick can live under such remarkable circumstances. As I have picked the extreme cases in the series given, it is possible that these extremes were experimental errors, and as in the Utah data, no information is given as what happened to the chicks, I have no proof that they did live. But from the large number of hatches that were recorded below 8% and above 15%, giving a variation of the actual water content in the chick's body of about 10%, it is evident that chicks do hatch under remarkable physiological difficulties. One explanation that suggests itself is, that as there is considerable variation in evaporation of individual eggs due to the amount of shell porosity, and the chicks that hatch in either ease may be the ones whose individual variations threw them nearer the normal.

By a further study from the Ontario data of the relation of the evaporation to the results in livable chicks, it can be readily observed that all good hatches have evaporation centering around the 12% moisture loss, and that all lots with evaporations above 15% hatch out extremely poor.

The general averages of the machines supplied with some form of moisture was 34% of all eggs set, in chicks alive at four weeks of age, while the machines ran dry gave only 20% of live chicks at a similar period.

Now, I wish to call attention to a further point in connection with evaporation. If the final measure of the loss of weight by evaporation were the only criterion of correct conditions of moisture in the chick's body, the hatches that show 12%, or whatever the correct amount of evaporation may be, should be decidedly superior to those en either side. That they are better has already been shown. But they are far from what they should he. An explanation is not hard to find. The correct content of moisture is not the only essential to the chick's well being at the moment of hatching, but during the whole period of incubation. Under our present system of incubation, the chick is immediately subject to the changing evaporation of American weather conditions. The data for that fact, picked at random, will be of interest. The following table gives the vapor pressure at Buffalo, N. Y., for twenty consecutive days in April:

1	170	11	342
2	130	12	286
3	95	13	219
4	103	14	248
5	110	16	212
6	106	16	198
7	154	17	241
8	183	18	306
9	245	19	261
10	311	20	204

Supposing a hatch to be started at the beginning of the above period, by the end of the first week, with the excessive evaporation, due to a low vapor pressure, the eggs would all be several percent below the normal water content; the fact that the next week was warm and rainy, and the vapor pressure rose until the loss was entirely counterbalanced, would not repair the injury, even though the eggs showed at the end of incubation exactly the correct amount of shrinkage. A man might thirst in the desert for a week, then, coming to a hole of water fall in and drown, but we would hardly accept the report of a normal water content found at the pest-mortem examination as evidence that his death was not connected with the moisture problem.

The change of evaporation, due to weather conditions, is, under hens, less marked than in incubators. This is because there are no drafts under the hen, and because the hen's moist body and the moist earth, if she sets on the ground, are separate sources of moisture which the changing humidity of the atmosphere does not affect. Among about forty hens set at different times at the Utah Station, the loss of moisture of which was determined at three equal periods of six days each, the greatest irregularity I found was as follows: 1st period, 5.81%; 2nd period, 3.86%: 3rd period, 6.15% Compare this with a similar incubator record at the same station in which the loss for the three periods was 5.63%, 9.18% and 2.15%.

I think the reader is now in position to appreciate the almost insurmountable difficulties in the proper control of evaporation with the common small incubator in our climate. It is little wonder that one of our best incubator manufacturers, after studying the proposition for some time, throw over the whole moisture proposition, and put out a machine in which drafts of air were slowed down by felt diaphragms and the use of moisture was strictly forbidden.

The moisture problem to the small incubator operator presents itself as follows: If left to the mercies of chance and the weather, the too great or too little evaporation from his eggs will yield hatches that will prove unprofitable. In order to regulate this evaporation, he must know and be able to control both vapor pressure and the currents of air that strike the eggs. Now he does not know the amount of vapor pressure and has no way of finding it out. The so-called humidity gauges on the market are practically worthless, and even were the readings on relative humidity accurately determined, they would be wholly confusing, for their effect of the same relative humidity on the evaporation will vary widely with variations of the out-of-door temperature.

If the operator knows or guesses that the humidity is too low, he can increase it by adding water to the room, or the egg chamber, but he cannot tell when he has too much, nor can he reduce the vapor pressure of the air on rainy days when nature gives him too much water. As to air currents he is little better off—he has no way to tell accurately as to the behavior of air in the egg chamber and changes in temperature of the heater or if the outside air will throw these currents all off, since they depend upon the draft principle.

Taking it all in all, the man with the small incubator had better follow the manufacturer's directions and trust to luck.

The writer has long been of the conviction that a plan which would keep the rate of evaporation within as narrow bounds as we now keep the temperature, would not only solve the problem of artificial incubation, but improve on nature and increase not only the numbers but the vitality or livability of the chicks. With a view of studying further the relations between the conditions of atmospheric vapor pressure, and the success of artificial incubation, I have investigated climatic reports and hatching records in the various sections of the world.

Figure 7 shows the average monthly vapor pressures at four points in which we are interested:

A study of the extent of daily variations is also of interest. As a general thing they are less extreme in localities when the seasonal variations are also less. In Cairo, however, which has a seasonal variation greater than San Francisco, the daily variations during the hatching season are much less than in California. This is due to a constant wind from sea to land, and an absolute absence of rainfall, conditions for which Egypt is noted.

Nearness to a coast does not mean uniform vapor pressure, for with wind alternating from sea to land, it means just the opposite.

As will be readily seen the months in spring which give the best hatches, occupy a median place in the humidity scale. The fact that both hens and machines succeed best in this period is to me very suggestive of the possibility that with an incubator absolutely controlling evaporation, much of the seasonal variation in the hatchability would disappear.

Month	Buffalo, N. Y.	St. Louis, Mo.	San Francisco	Cairo, Egypt
January	87	98	311	279
February	81	94	310	288
March	138	224	337	287
April	171	283	332	311
May	301	423	317	328
June	466	650	345	365
July	546	599	374	413
August	496	627	382	435
September	429	506	389	372
October	285	327	342	365
November	271	225	285	321
December	143	133	243	397

Figure 7. Annual variation in humidity for selected cities.

The uniform humidity of the California coast is shown in the above table. This is not inconsistent with the excellent results obtained at Petaluma.

The Egyptian hatcher in his long experience has learned just about how much airholes and smudge fire are necessary to get results. With these kept constant and the atmosphere constant, we have more nearly perfect conditions of incubation than are to be found anywhere else in the world, and I do not except the natural methods. The climatic conditions of Egypt cannot be equaled in any other climate, but as will be shown in the last section of this chapter, their effect can be duplicated readily enough by modern science and engineering.

Mr. Edward Brown, who was sent over here by the English government to investigate our poultry industry, was greatly surprised at our poor results in artificial incubation. Compared with our acknowledged records of less than 50% hatches, he quotes the results obtained in hatching 18,000 eggs at an English experiment station as 62%. I have not obtained any data of English humidity, but it is undoubtedly more uniform than the eastern United States.

VENTILATION—CARBON DIOXIDE

The last of the four life requisites we have to consider is that of oxygen. The chick in the shell, like a fish, breathes oxygen which is dissolved in a liquid. A special breathing organ is developed for the chick during its embryonic stages and floats in the white and absorbs the oxygen and gives off carbon dioxide. The amount of this breathing that occurs in the chick is at first insignificant, but increases with development. At no time, however, is it anywhere equal to that of the hatched chicks, for the physiological function to be maintained by the unhatched chicks requires little energy and little oxidation.

Upon the subject of ventilation in general, a great misunderstanding exists. Far be it for me to say anything that will cause either my readers or his chickens to sleep less in the fresh air, yet for the love of truth and for the simplification of the problem of incubation, the real facts about ventilation must be given.

In breathing, oxygen is absorbed and carbon dioxide and water vapor are given off. It is popularly held that abundance of fresh air is necessary to supply the oxygen for breathing and that carbon dioxide is a poison. Both are mistakes. The amount of oxygen normally in the air is about 20%. Of carbon dioxide there is normally three hundredths of one percent. During breathing these gasses are exchanged in about equal volume. A doubling or tripling of carbon dioxide was formerly thought to be "very dangerous." Now, if the carbon dioxide were increased 100 times, we would have only three percent, and have seventeen percent of oxygen remaining. This oxygen would still be of sufficient pressure to readily pass into the blood. We might breathe a little faster to make up for the lessened oxygen pressure. In fact such a condition of the air would not be unlike the effects of higher altitudes.

Some investigations recently conducted at the U. S. Experiment Station for human nutrition, have shown the utter misconception of the old idea of ventilation. The respiratory calorimeter is an air-tight compartment in which men are confined for a week or more at a time while studies are being made concerning heat and energy yielded by food products.

It being inconvenient to analyze such an immense volume of air as would be necessary to keep the room freshened according to conventional ventilation standards, experiments were made to see how vitiated the air could be made without causing ill effects to the subject.

This led to a remarkable series of experiments in which it was repeatedly demonstrated that a man could live and work for a week at a time without experiencing any ill effects whatever in an atmosphere of his own breath containing as high as 1.813% of carbon dioxide; or, in other words, the air had its impurity increased 62 times. This agrees with what every chemist and physiologist has long known, and that is that carbon dioxide is not poisonous, but is a harmless dilutant just as nitrogen. This does not mean that a man or animal may not die of suffocation, but that these are smothered, as they are drowned, by a real absence of oxygen, not poisoned by a fraction of 1% of carbon dioxide.

In the same series of experiments, search was made for the mysterious poisons of the breath which many who had learned of the actual harmlessness of carbon dioxide alleged to be the cause of the ill effects attributed to foul air. Without discussion, I will say that the investigators failed to find such poisons, but concluded that the sense of suffocation in an unventilated room is due not to carbon dioxide or other "poisonous" respiratory products, but is wholly due to warmth, water vapor, and the unpleasant odors given off by the body.

The subject of ventilation has always been a bone of contention in incubator discussions. With its little understood real importance, as shown in the previous section, and the greatly exaggerated popular notions of the importance of oxygen and imagined poisonous qualities of carbon dioxide, the confusion in the subject should cause little wonder.

106

A few years ago someone with an investigating mind decided to see if incubators were properly ventilated, and proceeded to make carbon dioxide determinations of the air under a hen and in an incubator. The air under the hen was found to contain the most of the obnoxious gas. Now, this information was disconcerting, for the hen had always been considered to be the source of all incubator wisdom. Clearly either the perfection of the hen or the conception of pure air must be sacrificed. Chemistry here came to the rescue, and said that carbon dioxide mixed with water formed an acid, and acid would dissolve the lime of an egg shell. Evidently the hen was sacrificing her own health by breathing impure air in order to soften up the shells a little so the chicks could get out.

Since it could have been demonstrated in a few hours in any laboratory, that carbon dioxide in the quantities involved, has no perceptible effect upon egg shells, it is with some apology that I mention that quite a deal of good brains has been spent upon the subject by two experiment stations. The data accumulated, of course, fails to prove the theory, but it is interesting as further evidence of the needlessness of the old fear of insufficient ventilation.

At the Ontario Station, the average amounts of carbon dioxide under a large number of hens was 0.22%, or about ten times that of fresh air, or one-sixth of that which the man breathed so happily in the respiratory calorimeter. With incubators, every conceivable scheme was tried to change the amount of carbon dioxide. In some, sour milk was placed which, in fermenting, gives off the gas in question. Others were supplied with buttermilk, presumably to familiarize the chickens with this article so they would recognize it in the fattening rations. In other machines, lamp fumes were run in, and to still others, pure carbon dioxide was supplied. The percentage of the gas present varied in the machines from 0.06% to 0.58%. The results, of course, vary as any run of hatches would. The detailed discussion of the hatches and their relation to the amount of carbon dioxide as given in Bulletin 160 of the Ontario Station, would be unfortunately confusing to the novice, but would make amusing reading for the old poultryman. Speaking of a comparison of two hatches, the writer, on page 53 of the bulletin says, "The increase in vitality of chicks from the combination of the carbon dioxide and moisture over moisture only, amounting, as it does, to 4.5% of the eggs set, seems directly due to the higher carbon dioxide content," I cannot refrain from suggesting that if my reader has two incubators, he might set up a prayer wheel in front of one and see if he cannot in like manner demonstrate the efficacy of Heavenly supplications in the hatching of chickens.

The practical bearing of the subject of ventilation in the small incubator is almost wholly one of evaporation. The majority of such machines are probably too much ventilated. In a large and properly constructed hatchery, such as is discussed in the last section of this chapter, the entire composition of the air, as well as its movement, is entirely under control. Nothing has yet been brought to light that indicates any particular attention need be given to the composition of such air save in regard to its moisture content, but as the control of this factor renders it necessary that the air be in a closed circuit, and not open to all outdoors, it will be very easy to subject the air to further changes such as the increasing oxygen, if such can be demonstrated to be desirable.

TURNING EGGS

The subject of turning eggs is another source of rather meaningless controversy. Of course, the hen moves her eggs around and in doing so turns them. Doubtless the reader, were he setting on a pile of doorknobs as big as his head, would do the same thing. As proof that eggs need turning, we are referred to the fact that yolks stick to the shell if the eggs are not turned. I have candled thousands of eggs and have yet to see a yolk stuck to the shell unless the egg contained foreign organisms or was several months old. However, I have seen hundreds of blood rings stuck to the shell. Whether the chick died because the blood rings stuck or whether the blood rings stuck because the chick died I know not, but I have a strong presumption that the latter explanation is correct, for I see no reason, if the live blood ring was in the habit of sticking to the shell, why this would not occur in a few hours as well as in a few days.

In the summer of 1901 I saw plenty of chicks hatched out in Kansas in egg cases, kitchen cupboards and other places where regular turning was entirely overlooked.

Mr. S. P. Collins, head of the Produce Department of Swift & Co., says that he was one time cruelly deserted in a Pullman smoker for telling the same story. The statement is true, however, in spite of Mr. Collins' unpleasant experience. Texas egg dealers frequently find hatched chicks in cases of eggs.

Upon the subject of turning eggs the writer will admit that he is doing what poultry writers as a class do on a great many occasions; i.e., expressing an opinion rather than giving the proven facts. In incubation practice it is highly desirable to change the position of eggs so that unevenness in temperature and evaporation will be balanced. When doing this it is easier to turn the eggs than not to turn them, and for this reason the writer has never gone to the trouble of thoroughly investigating the matter. But it has been abundantly proven that any particular pains in egg turning is a waste of time.

COOLING EGGS

The belief in the necessity of cooling eggs undoubtedly arose from the effort to follow closely and blindly in the footsteps of the hen, With this idea in mind the fact that the hen cooled her eggs occasionally led us to discover a theory which proved such cooling to be necessary. A more reasonable theory is that the hen cools the eggs from necessity, not from choice. In some species of birds the male relieves the female while the latter goes foraging.

But there is no need to argue the question. Eggs will hatch if cooled according to custom, but that they will hatch as well or better without the cooling is abundantly proven by the results in Egyptian incubators where no cooling whatever is practiced.

Searching for the "Open Sesame" of Incubation

The experiment station workers have, the last few years, gone a-hunting for the weak spot in artificial incubation. Some reference to this work has already been made in the sections on moisture and ventilation. Before leaving the subject I want to refer to two more efforts to find this key to the mystery of incubation and in the one case at least correct an erroneous impression that has been given out.

At the Ontario Station a patent disinfectant wash called "Zenoleum" was incidentally used to deodorize incubators, Now, for some reason, perhaps due to the belief that white diarrhoea was caused by a germ in the egg,[*] this idea of washing with Zenoleum was conceived to be a possible solution of the incubator problem. In the numerous experiments at that station in 1907 Zenoleum applied to the machine in various ways was combined with various other incipient panaceas and at the end of the season the results of the various combinations were duly tabulated. The machine with buttermilk and Zenoleum headed the list for livable chicks.

For reasons explained in the chapter on "Experiment Station Work," the idea of contrasting the results of one hatch of one sort with the average results of many hatches of another sort is very poor science. Feeling that the Station men would hardly be guilty of expressing as they did in favor of such a method without better reasons, I very carefully went over the results and compared all machines using Zenoleum to all machines without it. The results in favor of Zenoleum were less marked but still perceptible. I was somewhat puzzled, as I could see no rational explanation of the relation between disinfecting incubator walls and the hatchability of the chick in its germ-proof cage. Finally I hit upon the scheme of arranging the hatches by date and the explanation became at once apparent. The hatching experiments had extended from

---

[*] White diarrhoea is indeed caused by a bacterium, *salmonella pullorum*.

110

March to July, but the Zenoleum hatches were grouped in April and early in May, when, as one would expect from weather conditions, all hatches were running good, After allowing for this error Zenoleum appeared as harmless and meaningless as would the Attar of Roses.

The second link after the missing link of incubation to which I wish to call your attention also occurred at the Ontario Station. The latter case, however, is happier in that no unwarranted conclusions were drawn and that an interesting bit of scientific knowledge was added to the world's store. The conception to be tested was an offshoot from the carbon dioxide theory. You will remember at the Utah Station the idea was that carbon dioxide was to dissolve the shell so the chick could break out easier.

At the Guelph Station the conception was that the carbon dioxide might dissolve the lime of the shell for the chick to use in "makin' hisself." As an egg could not be analyzed fresh and then hatched, a number were analyzed from the same hens and others from those hens were then incubated with the various amounts of carbon dioxide, buttermilk, Zenoleum, and other factors. The lime content of the contents of the fresh egg averaged about 0.04 grams. At hatching time the lime in the chick's body averaged about 0.20 grams and was always several times as great as the maximum of the eggs.

Clearly calcium phosphate of the chick's bones is made by the digestion of the calcium carbonate from the shell and its combination with the phosphorus of the yolk. Certainly a remarkable and hitherto unexplained fact. The amount of lime required is not great enough, however, to materially weaken the shell, but, of course, the process is vital to the chick as bones are quite essential to his welfare, but it is an "inside affair" of which the three-tenths of one per cent of carbon dioxide incidentally present under the hen is entirely irrelevant.

A further observation made by the investigator is that the chicks which obtained the lowest amount of lime were abnormally weak. As long as we are powerless to aid the chick in digesting lime this fact, like the other, belongs in the field of pure, rather than applied science. I think that we are safe in saying that the weakness caused the shortage of lime rather than vice versa; if the writer remembers runts in other animals are usually a little short of bone material.

The chemist of the station is to be given special credit for not jumping at conclusions. In the summary of this work he states.' "There is apparently no connection between the amount of lime absorbed by the chick and the amount of carbon dioxide present during incubation."

THE BOX TYPE OF INCUBATOR IN ACTUAL USE

Although the fact is not so advertised and frequently not recognized even by the makers, the success of existing incubators is directly proportional to the extent with which they control evaporation. In order to show this I have only to call attention briefly to two or three of the most successful types of incubators on the market.

Let me first repeat that evaporation increases with increased air currents and with decreased vapor pressure. Now, the vapor pressure undergoes all manner of changes with the passing of storm centers and the changes of prevailing winds. But there is a general tendency for vapor pressure to increase with increase in outside temperature. Now, the movement of air in all common incubators depends upon the draft principle and the greater the difference in machine temperature and outside temperature the greater will be this draft. Thus, we have two factors combining to cause variation in the rate of evaporation. The tendency for the rate of airflow to vary is diminished when a cellar is used for an incubator room, but the cellar does not materially remedy the climatic variation in vapor pressure.

The general tendency of incubators as ordinarily constructed, is to dry out the eggs too rapidly. With a view of counteracting this, water is placed in pans in the egg room. A surface of water exposed to quiet air does not evaporate as fast as one might think, as is easily shown by the fact that air above rivers, lakes and even seas is frequently far from the saturation point. The result of the moisture pan with a given current of air is that the vapor pressure is increased a definite amount, but by no means is it regulated or made uniform. Inasmuch as too much shrinking is the most prevalent fault in box incubators, the use of moisture is on the whole beneficial, but in hot, murky weather, with less circulation and higher outside vapor pressure, the moisture is overdone and the operator condemns the system.

The subject not being clearly understood and no means being available for vapor pressure determinations, the system results in confusion and disputes. When the felt diaphragm machine was brought into the market it was advertised as a no-moisture machine. The result of the diaphragm is that of choking off air movement and consequently reducing evaporation. This gives exactly the same results as the use of moisture, but the machine is easier to operate and seemed to do away with the vexatious moisture problem which, together perhaps, with some fancied resemblance of felt diaphragms to hen feathers, has resulted in the widespread use of this type of machine.

The latest effort along the lines of reducing evaporation is the sand tray machine that followed in the wake of the Ontario investigation. This device simply gives a greater evaporating surface to the water and hence a greater addition to the vapor pressure. The results in practice I had given me by a man who last year hatched sixty-five thousand chicks and as many more ducklings.

He said: "The sand tray early in the season gave the best hatches and most vigorous chicks we had, but later on things got too wet and the chickens drowned." No nicer demonstration of science in practice could be desired.

In the present-day incubator of cither type we are wholly at the mercy of sudden climatic changes of vapor pressure. For the slower changes from season to season some control by greater and lesser amounts of supplied moisture or by ventilator slides is available, but little understood and seldom practiced.

It will certainly be of interest to my readers to know the actual hatches obtained with the prevailing type of box incubator. By actual hatches we mean the percent of live chicks taken out of the machine to the per cent of eggs put in. The ordinary published hatches, based on one per cent of fertile hatches, are a delusion and a snare. When eggs are tested out many dead germs come out with them and the separation of microscopic dead germs from the infertile egg is, of course, impossible. Such padded and "show" hatching records do not interest us.

Where incubators are run on top of the ground I have found the results to be poor, and to improve the bigger and deeper and damper and warmer and less ventilated the cellar is made. The reason for this is plain. In such a cellar the vapor pressure of the air is not only greater but is less influenced by the shifting vapor pressure of the outside air. In a good cellar the operator, though his knowledge of the factors with which he deals is grievously deficient, learns, through long and costly experience, about what addition of moisture or about what rate of ventilation will give him the best results. In a room more subject to outside influences, the conditions are so constantly changing that uniformity of practice never gives uniform results, and hence the operator is without guidance, either intelligent or blind, and the results are wholly a product of chance.

As proof of my contention I may give results of a series of full-season hatches for 1908, each involving several thousand eggs.

First, a state experiment station, the name of which I do not care to publish. Incubators kept in a cement basement which has flues in which fires were built to secure "ample ventilation." This caused a strong draft of cold, dry air, making the worst possible condition for incubation. The hatch for the season averaged 25% and was explained by lack of vitality in the stock.

Second, Ontario Agricultural College. A room above ground, moisture used in most machines and various other efforts being made to improve the hatches by a staff of half a dozen scientists. Results: Hatch 48%—incubator manufacturers call the experimenters names and say they are ignorant and prejudiced.

Third, Cornell University: dry ventilated basement representing typical conditions of common incubator practice of the country. Results: Hatch 52%, results when given out commonly based on fertile eggs and everyone generally pleased.

Fourth: One of the most successful poultrymen in New York State, who has, without knowing why, hit upon the plan of using a no-moisture type of incubator in a basement which is heated with steam pipes, which maintains temperature at 70° and has a cement floor which is kept covered with water. Results: Hatch 59%.

Fifth: As a fifth in such a series I might mention again the Egyptian machine with the uniform vapor pressure of the climate and the three chicks exchanged for four eggs.

While an official in the United States Department of Agriculture, I gathered data from original records of private plants covering the incubation of several hundred thousand eggs. Such information was furnished me in confidence as a public official and as a private citizen I have no right to publish that which would mean financial profit or loss to those concerned.

Of records where there were ten thousand or more eggs involved, the lowest I found was 44% and the highest that mentioned as the fourth case above, or 59%. The great majority of these records hung very closely around the 50% mark.

Figure 8 is a fair sample of such data. It is the record of hatching hen eggs for the first six months of 1908 at one of the largest poultry plants in America.

Month	Eggs Set	Chicks Hatched	Percent Hatched
January	4,213	1,585	38
February	6,275	2,339	34
March	17,900	6,903	38
April	18,819	10,265	55
May	24,458	14,438	59
June	13,100	6,614	55
Total	**84,855**	**42,234**	**50**

Figure 8. Seasonable hatchability.

THE FUTURE METHOD OF INCUBATION

The idea of the mammoth incubator which would hatch eggs by the hundred thousand and a minimum of expense is the dream of the American incubator inventor. We have long had available such methods of insulation and regulating the supply of heat as would point to the practicability of such a dream.

The past efforts in this direction have fallen down for the following simple reason: All eggs were placed in a single big room with a view of the man's entering the room to take care of them. Contact with cold walls, the opening of doors, the hatching of chicks or introduction of fresh eggs set up air currents, the hot air rising and the cold air settling until great differences in temperature would be found in the room. No systematic regulation of evaporation was contemplated, as the principles at stake or the means of such regulation were unknown.

The attempt just referred to was made several years ago by one of the most successful of incubator manufacturers and because of his failure other inventors were inclined to steer clear of the proposition. Meanwhile the need for such an incubator has grown enormously. At the time that above effort was made no duck ranch existed whose annual production ran over thirty or forty thousand ducklings, whereas we now have several in the one hundred thousand class.

Much more remarkable has been the growth of the day-old chick business. The discovery that newly hatched chicks could be successfully shipped hundreds of miles with less loss than shipping eggs for hatching, has resulted in a few years' time in the growth of hatcheries of considerable size where chicks are hatched by means of common incubators. Still another opportunity for the use of large hatcheries has been by the growth of poultry communities. There are other communities besides those mentioned in this book which would amply support public hatcheries. If half the poultry growers of Lancaster County, Pa., were to be prevailed upon to patronize a public hatchery, the county would support between fifteen and twenty 100,000 egg incubators. Any of the numerous trolley centers in Indiana, Ohio and

Southern Michigan would likewise be profitable locations for the establishment of public hatcheries.

The demand for the incubator of large capacity has, within the last year or so, brought two or three "mammoth" incubators into the market. The devices I now refer to consist of a row of box incubators which, instead of being heated by single lamps, are heated by continuous hot water pipes. This scheme effects a considerable saving in fuel cost and labor, but the bulkiness of construction and the woeful lack of evaporation control are still to be dealt with.

The writer now wishes briefly to describe the plan of construction and operation of a new type of hatchery, the success of which has recently been made feasible by inventions and technical knowledge hitherto unavailable. The plan of the hatchery is on that of a cold storage plant as far as insulation and general construction go. The eggs are kept in bulk in special cases which are turned as a whole and may rest on either of four sides. At hatching time the eggs are spread out in trays in a special hatching room, which is only large enough to accommodate chicks to the amount of One-sixth of the incubator capacity, for twice a week deliveries, or one-third if weekly deliveries are desired.

There are no pipes or other sources of heat in the egg chambers. All temperature regulation is by means of air heated (or cooled as the case may be) outside of the egg rooms and forced into the egg rooms by a motor driven cone fan, maintaining a steady current of air, the rate of movement of which may be varied at will. The air movement maintained will always be sufficiently brisk, however, to prevent an unevenness of temperature in different parts of the room.

So simple is this that the reader will doubtless wonder why it was not developed earlier. The reason is that air subject to the climatic influences will, with any forced draft sufficient to equalize temperature, result in a fatal rate of evaporation. Sprinkling the air has not generally been thought practical because of the notion that air must be used in the egg chamber but once, which involved quite a waste of heat necessary in warming a large bulk of air and evaporating sufficient water. Moreover, no means has, in the past, been available for making a sufficiently accurate measurement of the evaporating power of the air.

The hair hygrometers commonly sold to incubator operators are known by scientists to be absolutely unreliable. The range between the wet and dry bulb thermometers was found in the 1 Ontario experiments to give readings with and without fanning that varied 15 to 20% in relative humidity which, at the temperature of an egg chamber, would amount to a variation of three to four hundred of vapor pressure units, which, with the forced draft plan, would ruin a hatch of eggs in a few hours. The sling psychrometer as used by the U.S. Weather Bureau should, in the hands of an expert, give results making possible measurements accurate to two or three percent of relative humidity or forty to sixty units of vapor pressure. In contrast with these blundering instruments we now have available an instrument with which the writer has frequently determined vapor pressure accurately to within a range of two or three vapor pressure units and the instrument is capable of being constructed for even finer work.

As it is only by means of air with the moisture content absolutely controlled that the use of a large room becomes possible, we can now see why this type of hatching remained so long undeveloped. By means of such vapor pressure control the large egg chamber is not only feasible but the rate of evaporation at once becomes subject to the control of the operator and we achieve a perfection in artificial incubation hitherto unattained.

The means by which the air moisture is regulated is similar to that used in up-to-date cold storage plants where the air is made moist by sprinkling and dried with deliquescent salts. The regulation of vapor pressure, like that of temperature, may be by electrically moved dampers which switch a greater or less proportion of the incoming current to the sprinkler or dryer as the case may be. The ordinary incubator thermostat gives the necessary impulse for the control of the temperature dampers, while the instrument above referred to is used for the vapor pressure control.

As the entire air circuit is closed, the chemical composition of the air may also be regulated at will. This results in a reduction of the quantity of heat required to a minimum; in fact, with the incubator in full swing, the air will, at times, need cooling rather than warming.

The question of the cost of incubation by this method, or of profit of such a hatchery operated for the public is almost wholly one of the size of operations. Where sufficient eggs may be obtained and sufficient demand exists for the chicks to make it profitable to operate, the additional cost of hatching extra chicks will be insignificant compared with the present system.

The Egyptian poultryman gives four eggs for three chicks, but the American poultryman would be willing to give four eggs for one chick, as is shown by the fact that he sells eggs for from 1¢-3¢ apiece and buys day-old chicks for 10¢-15¢. A plant with a seasonable capacity of 100,000 eggs has a basis to work upon something as follows:

With a fifty percent hatch and chicks at 10¢ each there would be a gross income of $5,000 annually. From this we must subtract for eggs at 2¢ each, $2,000. Salary for operator $1,000, wages for helper $300. Fuel, supplies and repairs $500. Cost of delivery and sales of chicks $200. This leaves a residue of $1,000, which would pay a 20% interest on the necessary investment of $5,000. Personally. I think this is about the minimum unit of hatching that would prove worthwhile as an independent institution.

Any increase in the percentage of the hatch would, of course, reduce the unit of size necessary for profitable operation. Upon a single poultry plant, such as a duck farm, the cost of operation would be materially reduced, as the operator himself would take the place of the intelligent manager and the cost of gathering eggs and the delivery of the product would be eliminated.

The most profitable method of hatchery operation undoubtedly will be upon a plan analogous to what, in creamery operation, is called centralization. The success of this scheme depends upon the fact that transportation and agencies at country stores are relatively less important items of expense than plant construction and high salaries for skilled labor. A hatchery with a million capacity can be built and run at not more than twice the cost of one hundred-thousand plant and better men can be kept in charge of it. A portion of the saving will of course be expended in maintaining a system of buying eggs and selling chicks.

The material advantage of operating a hatchery in connection with a high-class egg handling and poultry packing establishment, or as one feature of a poultry community, is at once apparent, for the system of collecting the market produce will be utilized for gathering eggs and distributing chicks, each business helping the other.

The public hatchery also gives an excellent opportunity for the introduction of good stock among farmers who would be too shiftless to acquire it by ordinary methods.

Chapter 7. Feeding

The old adage that a little knowledge is a dangerous thing is nowhere better illustrated than in the scientific phases of poultry feeding. The attempted application of the common theoretical feeding standards to poultry has caused not only a great waste of time but has also resulted in expenditures for high-priced feeds when cheaper feeds would have given as good or better results.

The so-called science of food chemistry is really a rough approximation of things about which the actual facts are unknown. Such knowledge bears the same relation to accurate science as the maps of America drawn by the early explorers do to a modern atlas. Like these early efforts of geography the present science of food chemistry is all right if we realize its incompleteness. In practice, the poultryman, after a general glance at the "map," will find a more reliable guide in simpler things.*

I am writing this book for the poultryman, not the professor, and because I state that the particular kind of science wherein the professor has taken the most pains to teach the poultryman is comparatively useless, I fear it may arouse a mistrust of the value of science as a whole. I know of no way to prevent this except to point out the distinction between scientific facts and guesses couched in scientific language.

---

* When Hastings wrote this book, vitamins had yet to be discovered. In poultrykeeping, the main effect of this was that everyone knew chickens needed green feed and outdoor access, but no one knew why. Green feed provided all of the vitamins that were lacking in grain and beef scrap except vitamin D, which chickens, like humans, synthesize with the aid sunlight. Hastings was painfully aware of the gulf between theory and practice, and this is why he feels compelled to warn the reader against acting on advice that hasn't been proven in the field.

When a scientist states that a hen cannot lay egg shells containing calcium without having calcium in her food, that is a fact, and it works out in practice, for calcium is an element, and the hen cannot create elementary substances. When the same scientist, finding that an egg contains protein, says that wheat is a better egg food than corn because it has a larger amount of protein, that is a guess, and does not work in practice because protein is not a definite substance, but the name of a group of substances of which the scientist does not know the composition, and which may or may not be of equal use to the hen in the formation of eggs.

All substances of which the world is made are composed of elements which cannot be changed. When these elements are combined they form definite substances with definite proportions entirely independent of the original elements. The pure diamond is carbon. Gasoline is carbon and hydrogen. Several hundred other things are also carbon and hydrogen. Sugar is carbon combined with hydrogen and oxygen. These three elements make several thousand different substances, including fats, alcohol and formaldehyde. Hydrocyanic acid is carbon combined with hydrogen and nitrogen, and is the most deadly poison known.

The failure of food science is partly because we do not know the composition of many of the substances of food and partly because these substances are changed in the animal body in a manner which we do not understand and cannot control.

CONVENTIONAL FOOD CHEMISTRY

The conventional analysis of feeding stuff divides the food substances in water, carbohydrates, fat, protein and ash (minerals). The amount of water in the body is all-important, but, with the exception of eggs during incubation, I confess I prefer to rely upon the chicken's judgment as to the amount required.

The carbohydrate group contains starch, sugar, cellulose and a number of other things. Carbohydrates constitute two-thirds to three-fourths of all common rations and nine-tenths of that amount is starch. The proposition of how much carbohydrates the hen eats is chiefly determined by the quantity of grain she consumes.

Of fats there are many kinds of which the composition is definitely known. The amount of fats the hen eats is unimportant because she makes starch into fat. The protein or nitrogen-containing substances of the diet is the group of food substances over which most of the theories are expounded. The hen can make egg fat from corn starch or cabbage leaves because they contain the same elements. She cannot make egg white from starch or fat because the element of nitrogen, which is in the egg white, is lacking in the starch and fats.

The substances that have nitrogen in them are called proteins. They are very complex and difficult to analyze. In digestion these proteins are all torn to pieces and built up into other kinds of protein. Just as in tearing down an old house, only a portion of the material can be used in a new house, so it is with protein, and laboratory analysis cannot tell us how much of the old house can be utilized in building the new one.

In practice the whole subject simmers down to the proposition of finding out by direct experiment whether the hen will do the work best on this or that food, regardless of its nitrogen content as determined in the laboratory.

The results of many experiments and much experience has shown that lean meat protein will make egg protein and chicken flesh protein, and that vegetable protein pound for pound is not its equal. I know of no results that have proven that the high priced vegetable foods such as linseed meal, gluten feed, etc., have proven a more valuable chicken food than the cheapest grains.

With cows and pigeons this is not the case, but the hen is not a vegetarian by nature and high priced vegetable protein doesn't seem to be in her line. Of the three standard grains there is some indication of the value of the proteins for chickens are of the following ranks, 1st oats, 2nd corn, 3rd wheat.

The false conceptions of the value of wheat protein has been specially the cause of much waste of money. Digestive trials and direct experiments both show that, as chicken foods, wheat is worth less, pound for pound, than corn and yet, though much higher in price, it is still used not only as a variety grain, but by many poultrymen as the chief article of diet. Wheat contains only 3% more protein than corn. The man who substitutes wheat at one 1½¢ a pound for corn worth one cent a pound pays 17¢ a pound for his added protein. In beef scrap he could get the protein for 5¢ a pound and have a very superior article besides.

Milk as a source of protein ranks between the vegetable proteins and those of meat. It is preferably fed clabbered. The dried casein recently put on the market is a valuable food but is not worth as much as meat food and will not be extensively utilized until the demand for meat scrap forces up the price to a point where the casein can be sold more cheaply.

Meat scrap, to be relished by the chickens, must not be a fine meal, but should consist of particles the size of wheat kernels or larger. The fine scrap gives the manufacturer a chance to utilize dried blood and tankage which is cheaper in quality and price than particles of real meat.

The last and least understood of the groups of food substances is mineral substance or ash. Now, the chemist determines mineral substance by burning the food and analyzing the residue. In the intense heat numerous chemical changes take place and the substances that come out of the furnace are entirely different from these contained in the fresh food.

The lay reader will probably ask why the chemist does not analyze the substances of the fresh material. The answer is that he doesn't know how. Progress is made every year but the whole subject is yet too much clouded in obscurity to be of any practical application. At present the feeding of mineral substance, like the feeding of protein, can best be learned by experimenting directly with the foods rather than by attempting to go by their chemical composition.

In practice it is found that green feed supplies something which grain lacks, presumably mineral salts.[*] Moreover we know that such food fed fresh is superior to the same substance dried.

---

[*] In fact, it is vitamins.

This may be because of chemical changes that occur in curing or simply because of greater palatability.

The other chief source of mineral matter is meat preparations with or without ground bone. Recent experiments at Rhode island have attempted to show the relative value of the mineral constituents of meat by adding bone ash to vegetable proteins, as linseed and gluten meal. The results clearly indicate that mineral matter of animal origin greatly improves the value of the vegetable diet, but that the latter is still sadly deficient. Of course the burning process used in preparing the bone ash may have destroyed some of the valuable qualities of the mineral salts. Practically, we do not care whether the value of animal meal be due to protein, mineral salts or both.[*]

In time the world will become so thickly populated that we cannot afford to rear cattle and condemn a portion of the carcass to go through another life cycle before human consumption. By that time the necessary food salts will doubtless be known and we will be able to medicate our corn and alfalfa and do away with the beef scrap. The poultrymen will do well, however, not to count on the chemistry of the future, for the chemist that makes the "tissue salts" for the hen may manufacture human food with Niagara power and fresh eggs will come in tin cans.

---

[*] It turned out to be a matter of proteins, minerals, *and* vitamins—all of which are present in adequate amounts in beef scrap and bone meal but not in soybean meal and other plant proteins. Replacing beef scrap with soybeans requires careful fortification with a kaleidoscope of vitamins, minerals, and proteins. This is usually purchased in the form of a feed pre-mix (such as Fertrell's Nutri-Balancer), since creating such a beef-scrap replacer is quite beyond the individual farmer.

HOW THE HEN UNBALANCES BALANCED RATIONS

Let the poultryman who figures the nutritious ratio of chicken feed try this simple experiment. Place before a half dozen newly hatched chicks a feed of one of the commercial chick feeds. When they have had their fill, sacrifice these innocents on the altar of science and open their crops. He will find that one chick has eaten almost exclusively of millet seed, another has preferred cracked corn, another has filled up heavily on bits of beef scrap and mica crystal grit, while a fourth fancied oats and granulated bone. In short the chick has, in three minutes, unbalanced the balanced ration that it took a week to figure out. This experiment can be varied by placing hens in individual coops and setting before each weighed portions of every food in the poultry supply man's catalogue.

There is only one kind of feeding that will balance rations and that is to feed exclusively on wet mash. This is successfully done in the duck business, but the ways of the duck are not the ways of the more fastidious hen.

In dairy work the individual preferences of the cows are given attention and their whims catered to by the herdsman. I know of nothing that makes a man more feel his kinship to the beast than to hear a good dairyman talk of the personalities and preferences of his feminine cooperators.

With commercial chicken work, humanly guided individual feedings is out of the question, though, if used, it might hasten the coming of the two-egg-per-day hen. Individual feeding with the hen as sole judge as to what she shall eat, which means each food in separate hoppers and free range, is the best system of chicken feeding yet evolved.

The duty of the poultryman is to supply the food, giving enough variety to permit of the hens having a fair selection. In practice this means that every hen must have access to water, grit (preferably oyster shell), one kind of grain, one kind of meat, and one kind of green food. In practice it will pay to add granulated bone for growing stock. One or two extra grains for variety and as many green foods as convenient, to increase palatability—hence increase the amount of food consumed, for a heavy food consumption is necessary for egg production.

As corn is the cheapest food known, let it be the bread at the boarding house and other grains the rotating series of hash, beans and bacon. The grain hopper may have two divisions. The corn never changes but the other should have a change of grain occasionally. The extent of the use made of the various grains will be determined by their price per pound.

The proportions of food of the various classes that will be consumed is about as follows:

Of 100 lbs. of dry matter:

- 8 to 12 lbs. meat
- 66 to 75 lbs. grain
- 15 to 25 lbs. green food

The profits of the business will be increased by supplying the green food in such tempting forms as to increase the amount consumed and cut down the use of grains.

The methods we have been describing in which various dry unground grains, beef scrap and oyster shell, each in a separate compartment, are exposed before the hen at all times, together with the abundant use of green food, either as pasture or a soiling crop, is the method of feeding assumed throughout this book.

The hopper feeding of so-called dry mash or ground grain mixture has been quite a fad in the last few years. The tendency of the hens to waste such food has occasioned considerable trouble. They are picking it over for their favorite foods and trying to avoid disagreeable foods. This difficulty is relieved when the food is separated into its various components and the hen offered each separately. As a matter of fact, there is no occasion for feeding ground feed except in fattening rations and here the wet mash is desirable.

The use of the products of wheat milling has been the chief excuse for such practices, but unless these get considerably lower in price per pound than corn they may be left off the bill-or-fare to advantage. The great use made of these products in poultry feeding was chiefly a result of the attempted application of the balanced ration idea, but as has already been shown the efforts to raise the protein ratio with grain foods is generally false economy.*

The old-fashioned wet mash which the writer does not recommend because of the labor involved, is, nevertheless, a fairly profitable method of poultry feeding. It is used in the Little Compton district of Rhode Island and was also used in the famous Australian egg laying contests elsewhere described. Personally I would prefer feeding ground grain wet, especially wheat bran and middlings, to feeding it dry.

The scattering of grain in litter so generally recommended in poultry literature is all right and proper, but is rather out of place in commercial poultry farming. It is used on the large poultry plants with the yards and long houses, but is not used on colony farms or in any of the poultry growing communities. I should recommend littered houses for Section 6 and the northern half or Section 3 (see Chapter IV), but with warmer soils and climate where the snow does not lie on the ground it would add a labor expense that would very seriously handicap the business.

The systems of poultry feeding that are commonly advertised are based either on some patent nostrum or a recommendation of green food in novel form, such as sprouted oats. The joke about poultry feed at 10¢ a bushel, absurd though it may seem, bas caught lots of dollars. To take a bushel of oats worth 50¢, add water, let them sprout and have five bushels costing 10¢, is certainly a wonderful achievement in wealth getting. The only reason a man couldn't run a soup kitchen on the same principle is that he can't do a soup business by mail. Sprouted oats are a good green food, however, though somewhat laborious to prepare. I should

---

* Hens usually find beef scrap rather unpalatable, and don't eat as much as they should. Mixing it with grain can increase consumption. This is the main benefit of mash. As Hastings points out, the entrenched practice of bulking up the mash with wheat bran and other low-energy, unbalanced-protein feeds just makes the mash less palatable, largely defeating its purpose. This concept was suddenly rediscovered around 1940, when "high-energy diets" took over.

certainly recommend them if for any reason the regular green food supply should run out.

The points already mentioned are about all the practical suggestions that the science of animal nutrition has to offer the poultryman. The discussion of feeding from its technical viewpoint is sufficiently covered in the chapter on "Farm Poultry" and the discussion of the management and economics of various types of poultry production.

Chapter 8. Diseases

For the study of the classification and description of the numerous ailments by which individual fowls pass to their untimely end, I recommend any of the numerous books written upon the subject. Some of these works are more accurate than others, but that I consider immaterial. The study of these diseases is good for the poultryman; it gives his mind exercise. When a boy in high school I studied Latin for the same purpose.

DON'T DOCTOR CHICKENS.

For the cure of all poultry diseases when they have passed a point when the fowl does not eat or for other reasons recovery is improbable, I recommend a blow on the head—the hatchet spills the blood, which is unwise.

The usual formula of "burn or bury deeply" is somewhat troublesome, unless you have a furnace running. A covered pit is more convenient if far enough removed from the house that the odor is not prohibitive. A post with a tally card may be planted nearby. This part or the poultry farm may be marked "Exhibit A," and shown first to the visitor during the busy season. If he is one of those prospective pleasure-and-profit poultrymen who propose to disregard all facts of biology and economics of production, you may save yourself the trouble of showing him the rest of the plant. Unfortunately, this scheme is not open to the poultryman who has breeding stock for sale.

I have frequently had the question put to me in the smoker of a Pullman car, "Do not epidemic diseases make the poultry business precarious?" Such questions came from farm-raised men, but not from poultry farmers. Poultrymen should figure a certain loss of birds just as insurance companies figure on the human death rate, but to all practical intents and purposes the epidemic disease has been banished from the poultry farms and seldom if ever enters the records in an answer to the question on, "Why do poultry farms fail?" Some of my readers may take exception to me either in regard to roup or white diarrhoea. Roup is a disease of the wrong system and careless management. White diarrhoea, so called, is a matter of wrong incubation.

The high mortality of young chicks, though not an epidemic disease, shares with excessive cost of production, very much of the responsibility for poultry farm failures. At the present writing the poultry editors of the country are having much discussion over the conclusion of Dr. Morse of the Bureau of Animal Industry to the effect that white diarrhoea is caused by an intestinal parasite similar to the germ that causes human dysentery.[*] Dr. Morse's opportunities for investigation have been somewhat limited and as the intestines of any animal are always swarming with various organisms, it will take very conclusive evidence to prove that the doctor is right. Practically the naming of the germs that attend the funeral is not particularly important for the reason that it has been thoroughly demonstrated that with good parentage, good incubation and good brooder conditions, white diarrhoea is unknown.

THE CAUSES OF POULTRY DISEASES

Poultry ailments are assignable to one of the three following causes, or a combination of these: First, hereditary or inborn weakness; second, unfavorable conditions of food, surroundings, etc.; third, bacteria or animal parasites.

A great many chickens die while yet within the shell, or during the growing process, there being no assignable reason save that of inherited weakness. To this class of troubles the only remedy is to breed from better stock. It is as much the trait of some birds to produce infertile eggs or chicks of low vitality as it is for others to produce vigorous offspring.

---

[*] Actually, it is bacterial, caused by *salmonella pullorum*.

The second class of ailments needs no discussion save that accorded it under the general discussions of the method of conducting the business.

The third class of ailments includes the contagious diseases. It is now believed that most common diseases are caused by microscopic germs known as bacteria. These germs in some manner gain entrance to the body of an animal, and, growing within the tissues, give off poisonous substances known as toxins, which produce the symptoms of the disease. The ability to withstand disease germs varies with the particular animal and the kind of disease. As a general rule it may be stated that disease germs cannot live in the body of a perfectly vigorous and healthy animal. It is only when the vitality is at a low ebb, owing to unfavorable conditions or inherited weakness, that disease germs enter the body and produce disease.

The bacteria which cause disease, like other living organisms, may be killed by poisoning. Such poisons are known as disinfectants. If it were possible to kill the bacteria within the animal, the curing of disease would be a simple matter, but unfortunately the common chemical poisons that kill germs kill the animal also. The only thing that can be relied upon to kill disease germs within the animal, is a counter-poison developed by the animal itself and known as an anti-toxin. Such anti-toxins can be produced artificially and are used to combat certain diseases, as diphtheria and smallpox in human beings and blackleg in cattle. Such methods of combating poultry diseases have not been developed, and due to the small value of an individual fowl would probably not be commercially useful even if successful from a scientific standpoint. The only available method of fighting contagious diseases of poultry is to destroy the disease germs before they enter the fowls and to remove the causes which make the fowl susceptible to the disease.

Contagious diseases of poultry may be grouped into two general classes: First, those highly contagious; second, those contracted only by fowls that are in a weakened condition. To the first class belong the severe epidemics, of which chicken-cholera is the most destructive.

CHICKEN-CHOLERA

The European fowl-cholera has only been rarely identified in this country. Other diseases similar in symptoms and effect are confused with this. As the treatment should be similar the identification of the diseases is not essential.

Yellow or greenish-colored droppings, listless attitude, refusal of food and great thirst are the more readily observed symptoms. The disease runs a rapid course, death resulting in about three days. The death rate is very high. The disease is spread by droppings and dead birds, and through feed and water. To stamp out the disease, kill or burn or bury all sick chickens, and disinfect the premises frequently and thoroughly. A spray made of one-half gallon carbolic acid, one-half gallon of phenol and twenty gallons of water may be used. Corrosive sublimate, 1 part in 5000 parts of water, should be used as drinking water.* This is not to cure sick birds, but to prevent the disease from spreading by means of the drinking vessels. Food should be given in troughs arranged so that the chickens cannot infect the food with the feet. All this work must be done thoroughly, and even then considerable loss can be expected before the disease is stamped out. If cholera has a good start in a flock of chickens it will often be better to dispose of the entire flock than to combat the disease. Fortunately cholera epidemics are rare and in many localities have never been known.

---

* As a general rule, one should never use the medications or disinfectants recommended in old poultry books, as they are far more toxic than their modern equivalents, and often carcinogenic as well. For example, "corrosive sublimate" is mercuric chloride, and is intensely poisonous. It's not the sort of thing I would allow on my farm. This book was written in an era where people, even health-food pioneers like Hastings, were extremely casual about exposing themselves to poisons.

ROUP

This disease is a representative of that class of diseases which, while being caused by bacteria, can be considered more of a disease of conditions than of contagion. Roup may be caused by a number of different bacteria which are commonly found in the air and soil. When chickens catch cold these germs find lodgment in the nasal passages and roup ensues. The first symptoms of roup are those of an ordinary cold, but as the disease progresses a cheesy secretion appears in the head and throat. A wheezing or rattling sound is often produced by the breathing. The face and eyes swell, and in severe cases the chicken becomes blind. The most certain way of identifying roup is a characteristic sickening odor. The disease may last a week or a year. Birds occasionally recover, but are generally useless after having had roup.

Sick birds should be removed and destroyed, but the time usually spent in doctoring sick birds and disinfecting houses can in this case be better employed in finding and remedying the cause of the disease. Such causes may be looked for as dampness, exposure to cold winds, or to a sudden change in temperature as is experienced by chickens roosting in a tight house. Fall and winter are the seasons of roup, while it is poorly housed and poorly fed flocks that most commonly suffer from this disease. Flocks that have become thoroughly roupy should be disposed of and more vigorous birds secured. The open front house has proved to be the most practical scheme for the reduction of this disease.[*]

---

[*] Roup has several forms, the most common being "infectious coryza," a bacterial infection. Another form, "nutritional roup," is caused by vitamin A deficiency, which in Hastings' time was synonymous with a lack of palatable green feed. In most climates would occur in fall and winter. Both forms would occur together in damp, crowded houses where inadequate green feed was provided, which in Hastings' day was synonymous with bad management. I have never seen a case of roup.

CHICKEN-POX, GAPES, LIMBER NECK

Chicken-pox or sore-head is a disease peculiar to the South. It attacks growing chickens late in the summer. Southern poultry-men who give reasonable attention to their stock find that, while this disease is a source of some annoyance, the losses are not severe and that it may be readily controlled. In the first place, the animal epidemic of pox can be practically avoided by bringing the chicks out early in the season. If the disease does develop in the flock, the birds are taken from the coops at night and their heads dipped in a proper strength of one of the coal tar disinfectants. Such treatment once a week has generally been effective. This disease is an exception to the general rule that disinfectants which kill germs also kill the chicken. The explanation is that chicken-pox is an external disease.

Gapes is given in every poultry book as one of the prominent poultry diseases, but are not common in the Northern and Western States. Gapes are caused by a parasitic worm in the windpipe. Growing chicks are affected. The remedy is to move the chicks to fresh ground and cultivate the old.

Limber neck (botulism) is not a disease, but is the result of the fowl's eating maggots from dead carcasses. It can be prevented by not allowing dead carcasses to remain where the chickens will find them. No practical cure is known.

LICE AND MITES

The parasites referred to as chicken-lice include many different species, but in habit they may be classed as body-lice and roost-mites. The first, or true bird-lice, live on the body of the chicken and eat the feathers and skin. The roost-mite is similar to a spider and differs in habits from the body-louse in that it sucks the blood of the chicken and does not remain on the body of the fowl except at night.

Body-lice are to be found upon almost all chickens, as well as on many other kinds of birds. Their presence in small numbers on matured fowls is not a serious matter. When body-lice are abundant on sitting hens they go from the hen to the newly hatched chickens, and may cause the death of the chicks. The successful methods of destroying body-lice are three in number:

1. Dust or earth wallows in which the active hens will get rid of lice. Such dust baths should be especially provided for yarded chickens and during the winter. Dry earth can be stored for this purpose. Sitting hens should have access to dust baths.

2. The second method by which body-lice may be destroyed is the use of insect powder. The pyrethrum powder is considered the best for this purpose, but is expensive and difficult to procure in the pure state. Tobacco dust is also used. Insect powder is applied by holding the hen by the feet and working the dust thoroughly into the feathers, especially the fluff. The use of insect powder should be confined to sitting hens and fancy stock, as the cost and labor of applying is too great for use upon the common chicken.

3. The third method is suitable for young chicks, and consists of applying some oil and grease on the head and under the wings. Do not grease the chick all over.

With vigorous chickens and correct management the natural dust bath is all that is needed to combat the lice.

The roost-mite is probably the cause of more loss to farm poultry raisers than any other pest or disease. The great difficulty in destroying mites on many farms is that chickens are allowed to roost in too many places. If the chicken-house proper is the only building infected with mites the difficulty of destroying them is not great. Plainness in the interior furnishings of the chicken-house is also a great advantage when it comes to fighting mites. The mites in the daytime are to be found lodged in the cracks near the roosting-place of the chickens.

Mites can be killed with various liquids, the best in point of cheapness is boiling water. Give the chicken-house a thorough cleaning and scald by throwing dippers of hot water in all places where the mites can find lodgment. Hot water destroys the eggs as well as the mites. Whitewash is a good remedy, as it buries both mites and eggs beneath a coating of lime from which they cannot emerge. Pure kerosene or a solution of carbolic acid in kerosene, at the rate of a pint of acid to a gallon of oil, is an effective lice-paint. Another substance much used for destroying insects or similar pests is carbon bisulphide. This is a liquid which evaporates readily, the vapor destroying the insects or mites. Carbon bisulphide or other fumigating agents are not effective in the average chicken-house because the house cannot be tightly closed. The liquid lice-killers on the market are very effective. They are usually composed of the remedies just mentioned, or of something of similar properties.[*]

---

[*] Again, don't use the chemicals mentioned here without investigating them first. Pyrethrins are still used as an organic insecticide. I wouldn't touch carbolic acid, carbon bisulphide, or tobacco-based remedies with a ten-foot pole.

Chapter 9. Poultry Flesh And Poultry Fattening

The poultry flesh which is used for food may be grouped into three divisions:

1. Poultry carcasses grown especially for market.
2. Poultry carcasses consisting of hens and young male birds that are sold from the general farms where the pullets are kept for egg production.
3. The cockerels and old hens sold as a by-product from egg farms.

The third class hardly needs our consideration in the present chapter. This stock, usually Leghorns, like Jersey veal, is to be disposed of at whatever price the market offers.

The cockerel will, if growing nicely, be fairly plump, and the hens, if on hopper rations of corn and beef scrap, will be about as fat as they can be profitably made, and to waste further effort upon them would not pay. Leghorn cockerels and hens are a wholesome enough meat, but will never command fancy prices nor warrant extra pains.

In class two we find the great mass of the poultry flesh of the country. This stock consisting chiefly, as it does, of Plymouth Rocks and Wyandottes, is well worth some extra pains toward increasing its quantity and quality.

Within the last ten or fifteen years several changes have been brought about in the general methods of handling farm poultry. Formerly it was thought desirable to market all stock not kept as layers while in the broiler stage of from 1½ to 2 pounds. Since the introduction of the custom of holding fall broilers over in cold storage, the price has fallen until it is now more profitable to market the surplus cockerels from the farm at three or four months of age. At this period the flesh has cost less per pound to produce than at either an earlier or later stage. For such purposes only the well-fleshed type of American breeds has been found desirable. The Leghorns and similar breeds are too small and become staggy too soon.

Contrary to a common belief and to the custom in the poultry books of classifying the Asiatics as "meat breeds," the Brahmas and Cochins are among the very poorest fowls that can be used for farm production of poultry meat. At the age spoken of, these breeds are lanky and unsightly and not wanted by poultry packers.

Consecutively with and perhaps responsible for change of sentiment that demands that broilers be allowed to grow into four-pound chickens, we find the development of the crate fattening industry.

CRATE-FATTENING

The introduction of crate-fattening into the Central West occurred about 1900. The credit of this introduction belongs to the large meat packing firms. At the present time the business is not confined to the meat packers, but is shared by independent plants throughout the country.[*]

The plants of the West range from a few hundred to as high as 20,000 capacity. They are constructed for convenience and a saving of labor, and in this respect are decidedly in advance of the European establishments where fattening has been long practiced.

The room used for fattening is well built and sanitary. A good system of ventilation is essential, as murky, damp air breeds colds and roup. The coops are built back to back, and two or more coops in height. Each coop is high and wide enough to comfortably accommodate the chickens, and long enough to contain from five to twelve chickens. The chickens stand on slats, beneath which are dropping-boards that may be drawn out for cleaning. The dropping-boards and feeding-troughs are often made of

---

[*] Crate-fattening is no longer practiced in this country. However, I have heard it said that we don't know what we're missing; milk-fattened chicken has a texture and flavor unequalled by any other kind of poultry.

metal. Strict cleanliness is enforced. No droppings or feed are allowed to accumulate and decompose.

As is a general rule in meat production, young animals give much better returns for food consumed than do mature individuals. With the young chicken the weight is added as flesh, while the hen has a tendency, which increases with age, to turn the same food into useless fat. For this reason the general practice is to fatten only the best of the young chickens. The head feeder at a large and successful poultry plant gave the following information on the selection of birds for the fattening-crates:

"The younger the stock the more profitable the gain. All specimens showing the slightest indication of disease are discarded. The Plymouth Rock is the favorite breed, and the Wyandotte is second. Leghorns are comparatively fat when received, and, while they do well under feed and 'yellow up' nicely, they do not gain as much as the American breeds. Black chickens are not fed at all. Brahmas and Cochins are not considered good feeders at the age when they are commonly sold. Chickens in fair flesh at the start make better gains than those that are extremely lean or very fat. But, contrary to what the amateur might assume, the moderately fat chicken will continue to make fair gains, while the very lean chicken seldom returns a profit."

The idea has been somewhat prevalent that there is some guarded secret about the rations used in crate-fattening. This is a mistaken notion. The rations used contain no new or wonderful constituent, and although individual feeders may have their own formulas, the general composition of the feed is common knowledge. The feed most commonly used consists of finely ground grain, mixed to a batter with buttermilk or sour skim-milk. The favorite grain for the purpose is oats finely ground and the hulls removed. Oats may be used as the sole grain, and is the only grain recommended as suitable to be fed alone. Corn is used, but not by itself. Shorts, ground barley or ground buckwheat are sometimes used. Beans, peas, linseed and gluten meals may be used in small quantities. When milk products are obtainable they are a great aid to successful fattening. Tallow is often used in small quantities toward the finish of the feeding period. The assumption is that it causes the deposit of fat globules throughout the muscular tissues, thus adding to the quality of the meat. The following simple rations show that there is nothing complex about the crate-fed chicken's bill of fare:

No. 1.—Ground oats, 2 parts; ground barley, 1 part; ground corn, 1 part; mixed with skim-milk.

No. 2.—Ground corn, 4 parts; ground peas, 1 part; ground oats, 1 part; meat-meal, 1 part; mixed with water.

A ration used by some fatters with great success is composed of simply oatmeal and buttermilk.

The feed is given as a soft batter and is left in the troughs for about thirty minutes, when the residue is removed. Chickens are generally fed three times per day. Water may or may not be given, according to the weather and the amount of liquid used in the food.

The chicken that has been crate-fattened has practically the same amount of skeleton and offal as the unfattened specimen, but carries one or two pounds more of edible meat upon its carcass. Not only is the weight of the chicken and amount of edible meat increased, but the quality of the meat is greatly improved, consisting of juicy, tender flesh. For this reason the crate-feeding process is often spoken of as fleshing rather than as fattening.

The enforced idleness causes the muscular tissue to become tender and filled with stored nutriment. The fatness of a young chicken, crate-fed on buttermilk and oatmeal, is a radically different thing from the fatness of an old hen that has been ranging around the corn-crib.

The crate-fattening industry while deserving credit for great improvement in the quality of chicken flesh in the regions where it has been introduced, cannot on the whole be considered a great success. It is commonly reported that some of the firms instrumental in its introduction lost money on the deal. The crate-fattening plant has come to stay in the communities where careful methods of poultry raising are practiced, and where the stock is of the best, but when a plant is located in a newly settled region where the poultry stock is small and feed scarce, the venture is pretty apt to prove a fiasco.

While poultryman at the Kansas Experiment Station, the writer made a large number of individual weighings of fowls in the crates of one of the large fattening plants of the state.

These weighings pointed out very clearly why the expected profits had not been realized. The birds selected for weighing were all fine, uniform looking Barred Rock Cockerels. At the end of the first week they were found to still appear much the same, but when handled a difference was easily noticed. By the end of the second week a few birds had died and many others were in a bad way. The individual changes of weight ran from 2½ pounds gain to ¾ pound loss, and many of the lighter birds were of very poor appearance. It is simply a matter of forced feeding being a process that makes trouble with the health of the chicken if all is not just right.

It is probable that in the future more fattening will be done on the farm, or by the farmer operating in a small way among his neighbors. The reason for this is that the saving of labor in the large plant is hardly as great as the added loss from the shrinkage of the birds due to the excitement of shipping and crowding, and the introduction of disease by the mingling of chickens from so many different sources.

The Canadians especially have encouraged fattening on the farm. The following is a hand-bill gotten out by an enterprising Canadian dealer for distribution among the farmers of his locality:

How To Fatten Chickens For The Export Trade

To fatten birds for the export trade, it is necessary to have proper coops to put them in. These should be two feet long, twenty inches high and twenty inches deep, the top, bottom and front made of slats. This size will hold four birds, but the cheapest plan is to build the coops ten feet long and divide them into five sections.

What to Feed

Oats chopped fine, the coarse hulls sifted out, two parts; ground buckwheat, one part; mix with skim-milk to a good soft batter, and feed three times a day. Or, black barley and oats, two parts oats to one part barley. Give clean drinking water twice a day, grit twice a week, and charcoal once a week. During the first week the birds are in the coops they should be fed sparingly—only about one-half of what they will eat. After that gradually increase the amount until you find out just how much they will eat up clean each time. Never leave any food in the troughs, as it will sour and cause trouble. Mix the food always one feed ahead. Birds fed in this way will be ready for the export trade in from four to five weeks. Chickens make the best gain put in the coop weighing three to four pounds.

We Supply the Coops

We have on hand a number of coops for fattening chicks, which we will loan to any person, "free of charge", who will sign an agreement to bring all chicks fattened in them to us. Every farmer should have at least one of these coops, as this is the only way to fatten chicks properly. In this way you can get the highest market price. We can handle any quantity of chicks properly fatted

ARMSTRONG BROS.

The farmer who does not think it worth while to construct fattening crates for his own crop of chickens may get very fair results by simply enclosing the chickens in some vacant shed. To these may feed a ration of two-thirds corn meal and one-third shorts, or some of the more complicated rations used at the fattening plants may be fed.

In the East, poultry fattening on the general farm is not dissimilar from the practices in the Central West, but we find a larger use of cramming machines, caponizing, and the growing of chickens for meat as an industry independent of keeping hens for egg production.

The cramming machine is a device by means of which food in a semi-liquid state is pumped into the bird's crop, through a tube inserted in the mouth. This means of feeding is much more used in Europe than in this country. It requires good stock and careful workmen. The method will probably slowly gain ground in this country. The feed used in cramming is similar to that used in ordinary crate feeding, except that it is mixed as a thin batter.

CAPONIZING

Caponizing is the castration of male chickens. Capons hold the same place in the poultry market as do steers in the beef market.

Caponizing is practiced to quite an extent in France, and to a less degree in England and the United States.[*]

Much the larger part of the industry is confined to that portion of the United States east of Philadelphia, though increasing numbers of capons are being raised in the North Central States. During the winter months capon is regularly quoted in the markets of the larger eastern cities. Massachusetts and New Jersey are the great centers for the growing of capons, while Boston, New York, and Philadelphia are the great markets. In many eastern markets the prices paid for dressed capons range from 20¢ to 30¢ a pound. The highest prices usually prevail from January to May, and the larger the birds the more they bring a pound.

The purpose of caponizing is not, as is sometimes stated, to increase the size of the chicken, but to improve the quality of the meat. The capon fattens more readily and economically than other birds. As they do not interfere with or worry one another, large flocks may be kept together.

The breeds suitable for caponizing are the Asiatics and the Americans. Brahmas will produce, with proper care and sufficient time, the largest and finest capons. On the ordinary farm, where capons would be allowed to run loose, Plymouth Rocks would prove more profitable. Plymouth Rocks, Brahmas, Langshans, Wyandottes, Indian Games, may all be used for capons. Leghorns are not to be considered for this purpose.

Capons should be operated upon when they are about ten weeks or three months old and weigh about two pounds.

The operation of caponizing is performed by cutting in between the last two ribs. Both testicles may be removed from one

---

[*] Caponizing is no longer practiced on a large scale. However, gourmets still pay high prices for capons, and the special tools and instruction booklets are still being produced.

side or both sides may be opened. The cockerel should be starved for twenty-four hours in order to empty the intestines. Asiatics are more difficult to operate on than Americans, the testicles being larger and less firm. There is always some danger of causing death by tearing blood vessels, but the percent of loss with an experienced operator is very small. Loss by inflammation is still more rare. The testicle of a bird is not as highly developed as in a mammal, and if the organ is broken and a small fragment remains attached it will produce birds known as "slips." Some growers advise looking over the capons and puncturing the wind puffs that gather beneath the skin. This, however, is not necessary.

A good set of tools is indispensable and can be purchased for from $2 to $3. As a complete set of instructions is furnished with each set it is unnecessary to go into details here. The beginner should, however, operate on several dead cockerels before attempting to operate on a live one.

After caponizing the bird should be given plenty of soft feed and water. The capon begins to eat almost immediately after the operation is performed, and no one would suppose that a radical change had taken place in his nature.

The feeding of capons differs little from the feeding of other growing chickens. Corn, wheat, barley and sorghum would be suitable grain, while beef-scrap would be necessary to produce the best growth.

About three weeks before marketing place the capons in small yards and feed them three or four times a day, giving plenty of corn and other feed, or fatten them in one of the ways indicated in the section on fattening poultry. Corn meal and ground oats, equal parts by weight, moistened with water or milk, make a good mash for fattening capons.

In dressing capons leave the head and hackle feathers, the feathers on the wings to the second joint, the tail feathers, including those a little way up the back, and the feathers on the legs halfway up to the thigh. These feathers serve to distinguish capons from other fowls in the market. Do not cut the head off, for this is also a distinguishing feature of the capon, on account of the undeveloped comb and wattles.

The price received for capons is greater than any other kind of poultry meat except early broilers. There may be trouble in some localities in getting dealers to recognize capons as such and pay an advanced price.

On several farms in Massachusetts, 500 to 1,000 capons are raised annually, and on one farm 5,000 cockerels are held for caponizing. The industry is growing rapidly year by year and the supply does not equal the demand.

It is to be expected that the amount of caponizing done in the West will gradually increase. Those wishing to try the growing of capons will do well to secure an experienced operator. Good men at this work receive 5¢ per bird. Poor operators are dear at any price, as they produce a large number of worthless slips.

Chapter 10. Marketing Poultry Carcasses

In the marketing of poultry carcasses as in other phases of the industry, we really have two systems to discuss. The one is used for the marketing of the product of the farm of the Central West, and the other the product of the poultryman or eastern farmer, who is near a large market and who will be repaid for taking special pains in preparing his poultry for market.

FARM-GROWN CHICKENS

At the present time almost the entire poultry crop of the Central West is sold from the farm as live poultry. This farm stock is purchased by produce buyers or general merchants and shipped to the nearest county seat or other important town, where there are usually one or more poultry-killing establishments. These establishments may vary from a simple shed, where the chickens are picked and packed in barrels, to the more modern poultry-packing establishment, with its accommodations for fattening, dressing, packing, freezing, and storing.

The poultry-buying stations may be branches of the larger packing establishments, branch houses of large produce firms, or small firms operating independently and selling in the open market.*

---

* The practice of buying live chickens from anyone who had them to sell, then butchering them for the general marketplace, has entirely vanished. It has been replaced by the vertically integrated broiler industry. Some of Hastings' comments about quality will give some indication why this happened. The average farm chicken in this country was never really a very good chicken, though the best farm chickens were superb.

The chickens as purchased are grouped into the following classes: Springs ("spring chickens" or young broilers), hens, old roosters and (at certain seasons) young roosters or staggy cockerels. Early in the season small springs are quoted as broilers, while capons form a separate item where such are grown.

Chickens are starved before killing, for the purpose of emptying the crop, and, to some degree, the intestines. If this is not done the carcass presents an unsightly appearance and spoils more readily in storage.

The method of picking (plucking) is not always the same, even in the same plant. Scalding is frequently used for local trade, in the summer season, or with cheap-grade stuff. The greater portion of the stock is picked dry. The pickers are generally paid so much per bird. In some plants men do the roughing while girls are employed as pinners. Pickers work either with the chickens suspended by a cord or fastened upon a bench adopted to this purpose. The killing is done by bleeding and sticking. The last thrust reaches the brain and paralyzes the bird. The manner of making these cuts must be learned by practical instruction.* The feathers are saved, and amount to a considerable item. White feathers are worth more than others. The head and feet are left on the chicken and the entrails are not removed.

The bird, after being chilled in ice-water or in the cooling room, is ready for grading and packing. This, from the producer's standpoint, is the most interesting stage in the process, for it is here that the quality of the stock is to be observed. The grading is made on three considerations: (1) The general division of cocks, springs, hens and capons is kept separate from the killing-room; (2) the grading for quality; (3) the assortment according to size.

The grading for quality depends on the general shape of the chicken, the plumpness or covering of meat, the neatness of picking, the color of skin and legs, and the appearance of the feet and head, which latter points indicate the age and condition of health.

---

* The practice of sticking and dry-plucking is something of a lost art. These days, almost all chickens are scalded and wet-plucked.

The culls consist of deformed and scrawny chickens. The seconds are poor in flesh, or they may be, in the case of hens, unsightly from over-fatness. They are packed in barrels and go to the cheapest trade. Those carcasses slightly bruised or torn in dressing also go in this class. Although a preference is generally stated for yellow-skinned poultry, the white-skinned birds, if equal in other points, are not underranked in this score. The skin color that is decidedly objectionable is the purplish tinge, which is a sign of diseased stock. Black pin-feathers and dark-colored legs are a source of objection. This is especially true with young birds that show the pin-feathers. Feathered legs are slightly more objectionable than smooth legs. Small combs and the absence of spurs give better appearance to the carcass.

The following is the nomenclature and corresponding weights of the farm marketed chickens. In each class there will be seconds and culls. The seconds of each group are kept separate, but not graded so strictly; or perhaps not graded at all for size. The culls are packed in barrels and all kinds of chickens from fryers to old roosters here sojourn together until they reach their final destination, as potted chicken or chicken soup.

Broilers[*]—Packed in two weights. 1st: Less than two pounds; 2nd: between 2 and 2½ pounds.

Chickens[†]—Packed in three weights. 1st: between 2½ and 3 pounds; 2nd: between 3 and 3½ pounds; 3rd: between 3½ and 4 pounds.

Roasters[‡]—Packed in two weights. 1st: between 4 and 5 pounds; 2nd: above 5 pounds.

Stag Roosters—Cockerels, showing spurs and hard blue meat, packed in two weights. 1st: under 4 pounds; 2nd: above 4 pounds.

---

[*] These have no modern equivalent; think of them as skinny Cornish Game Hens.

[†] We would call these "broilers." They are small by modern standards.

[‡] These are about the same size as modern broilers, but are older, more flavorful, and tougher.

Fowls—hens. They are packed in three sizes. 1st: under 3¼ pounds; 2nd: between 3¼ and 4½ pounds; 3rd: over 4½ pounds.

Old Roosters—Packed in barrels. One grade only.

After packing, chickens may be shipped to market immediately, or they may be frozen and stored in the local plant. Shipments of any importance are made in refrigerator cars.

The poultry that is shipped to the final market alive is gradually diminishing in quantity as poultry killing plants are built up throughout the country. The live poultry shipments are chiefly made in special Live Poultry Transportation Cars. Figure 9 gives the number of such cars that moved out of the States named in a recent year.

Iowa	645	Tennessee	189
Missouri	630	Michigan	165
Illinois	624	S. Dakota	103
Kentucky	472	Oklahoma	101
Nebraska	395	Indiana	100
Kansas	370	Wisconsin	93
Minnesota	174	Texas	91
Ohio	173	Arkansas	47

Figure 9. Shipments from selected states, carload lots

The most of this live poultry goes to New York and other eastern cities and is consumed largely by the Jewish trade.

THE SPECIAL POULTRY PLANT

The special egg farmer of the East should sell his poultry alive to the regular dealer. The exception to this advice may he taken in the case of squab broilers (small Leghorn broilers) for which some local dealers will not pay as fancy a price as may be obtained by dressing and shipping to the hotel trade.

The grower of roasters and capons will probably want to market his own product. As to whether it will pay him to do so will depend upon whether his dealer will pay what the quality of the goods really demands. The dealer can afford to do this if he will hustle around and find an outlet for the particular grade of goods, for he is in position to kill and dress the fowls more economically than the producer.

I have never been able to study out why the average writer upon agricultural subjects is always advising the farmer to attempt to do difficult work for which special firms already exist. In the case of fattening just referred to, there is reason why the farmer may be able to do the work more successfully than the special establishment, but why any one should urge the farmer to turn the woodshed into a temporary poultry packing establishment I can hardly see. If the farmer has nothing to do he had better get a job at the poultry killing house where they have ice water and barrels in which to put the feathers.[*]

I do not think it worth while in this book for me to attempt to describe in detail the various methods of killing and packing poultry for the various retail markets. The grower who contemplates killing his own stuff had better spend a day visiting the produce houses and market stalls and inquire which methods are locally in demand.

Suggestions from Other Countries

In European countries generally, and especially in France and England, great pains are taken in the production of market poultry. Each farmer and each neighborhood became known in the market for the quality of their poultry, and the prices they receive

---

[*] Although it is the fashion these days for small producers to butcher their own chickens, this is almost entirely due to lack of establishments willing to do this for them. Learning to butcher chickens efficiently is time-consuming, and the facilities and equipment are expensive.

vary accordingly. In these countries more poultry is fattened and dressed by the growers than in the United States where we have greater specialization of labor.

In countries that have an export trade different systems have originated. In Denmark and Ireland cooperative societies are organized to handle perishable farm products. These, however, deal more with eggs than with poultry. In portions of England the fattening is done by private fatteners. The country being thickly settled, the chickens are collected directly from the farms by wagons making regular trips. This allows the rejection of the poor and immature specimens, whereas a premium may be paid on better stock.

The greatest fault of poultry buying as conducted in this country is the evil of a uniform price. After chickens are plucked the difference of quality is readily discerned, and the price varies from fancy quotations to almost nothing for culls. But the packer pays the farmer a given rate per pound for live hens or for spring chickens. The price is paid alike for the best poultry received or for the scrawniest chickens that can be coaxed to stand up and be weighed. The price paid is the average worth of all chickens purchased at that market. All farmers who market an article better than the average are unjust losers, while those who sell inferior stock receive unearned profits. The producer of good stock receives pay for the extra quantity of his chickens, but for the extra quality no recognition whatever is given. To the deserving producer, if quality were recognized, it would result in a greatly increased stimulation of the production of good poultry.

Any packer, if questioned, will state that he would be willing to grade chickens and pay for them according to quality, but that he does not do so because his competitor would pay a uniform price and drive him out of business. The man who receives an increased price would say little of it, while the man who sells poor chickens, if he failed to receive the full amount to which he is accustomed, would think himself unjustly treated and use his influence against the dealer. A recognition of quality in buying is for the interest of both the farmer and the poultry dealer, and a mutual effort on the part of those interested to put in practice this reform would result in a great improvement of the poultry industry.

COLD STORAGE OF POULTRY

The growth of the cold storage of poultry has been phenomenal. Poultry is packed in thin boxes that will readily lose their heat and these are stacked in a freezer with a temperature near the zero point. The temperature used for holding poultry are anywhere from 0° up to 20°. Poultry is held for periods of one to six weeks at temperature above the freezing point.

Frozen poultry will keep almost indefinitely save for the drying out, which is due to the fact that evaporation will proceed slowly even from a frozen body. The time frozen poultry is stored varies from a few weeks to eight or ten months.

The usual rule is that any crop is highest in price when it first comes on the market and cheapest just after the point of its greatest production. Thus, broilers are high in May and cheap in September. In such cases the goods are carried from the season of plenty to the following season of scarcity. This period is always less than a year. The idea circulated by wild writers, that cold storage poultry was kept several years, is an economic impossibility. The interest on the investment alone would make the holding of storage goods into the second season of plenty quite unprofitable, but when the costs of storage, insurance and shrinkage are to be paid, storing poultry for more than one season becomes absurd.

The fowl that has been once frozen cannot be made to look "fresh killed" again. For that reason packers like to get a monopoly on a particular market so that the two classes of goods will not have to compete side by side. The quality of the frozen fowl when served is very fair, practically as good as and some say better than the fresh killed.

Cold storage poultry is best thawed out by being placed overnight in a tank of water. Poultry prejudice prevents the practice of retailing the goods frozen, though this method would be highly desirable.

DRAWN OR UNDRAWN FOWLS

Within the last two or three years there has been a great hue and cry about the marketing of poultry without drawing the entrails.

The objection to the custom rests upon the general prejudice to allowing the entrails of animals to remain in the carcass. If a little thought is given the subject, however, it is seen that human prejudice is very inconsistent in such matters. We draw beef and mutton carcasses, to he sure, but fish and game are stored undrawn, and as for oysters and lobsters we not only store them undrawn but we eat them so.

The facts about the undrawn poultry proposition are as follows: The intestines of the fowl at death contain numerous species of bacteria, whereas the flesh is quite free from germs. If the carcass is not drawn, but immediately frozen hard, the bacteria remain inactive and no essential change occurs. If the carcass is stored without freezing, or remains for even a short time at a high temperature, the bacteria will begin to grow through the intestinal walls and contaminate the flesh.

Now, if the fowl is drawn, the unprotected flesh is exposed to bacterial contamination, which results in decomposition more rapidly than through the intestinal walls. The opening of the carcass also allows a greater drying out and shrinkage.

If poultry carcasses were split wide open as with beef or mutton, drawing might not prove as satisfactory as the present method, but since this is not desirable, and since ordinary laborers will break the intestines and spill their contents over the flesh, and otherwise mutilate the fowl, all those who have had actual experience in the matter agree that drawing poultry is unpractical and undesirable.[*]

As far as danger of disease or ptomaine poison is concerned, chances between the two methods seem to offer little choice.

---

[*] The emerging broiler industry spent pretty much a whole decade—the Fifties—convincing consumers that "ready to cook" eviscerated broilers were being prepared in a sanitary fashion.

The Bureau of Chemistry of the U. S. Department of Agriculture has conducted a series of experiments along the line of poultry storage. So far as the results have been published, nothing very striking has been learned. From what has been published, the writer is of the opinion that the somewhat mysterious changes that were observed in the cold storage poultry were mostly a matter of drying out of the carcass.

POULTRY INSPECTION

The enthusiastic members of the medical profession, and others whose knowledge of practical affairs is somewhat limited, occasionally come forth with the idea of an inspection of poultry carcasses similar to the Federal inspection of the heavier meats.

The reasons that are supposed to warrant the Federal meat inspection are precaution against disease and the idea of enforcing a cleanliness in the handling of food behind the consumer's back, which he would insist upon were he the preparer of his own food products.

No doubt there is well established evidence that some diseases, such as the dread trichinosis, are acquired by the consumption of diseased meat. As far as it is at present known there are no diseases acquired from the consumption of diseased poultry flesh, but, as we do not know as much about the bacteria that infests poultry as we do of that of larger animals, there is no positive proof that such transmission of disease could not occur. Thorough cooking kills all disease germs, and poultry is seldom, if ever, eaten without such preparation.

The idea of protecting people from uncleanly methods of handling their foods, concerning which they cannot themselves know, is somewhat of a sentimental proposition. In practice it amounts to nothing, save as the popular conception of this protection increases the demand for the product which is marked "U.S. Inspected and Passed."

It may be interesting to some of the reformers of 1906 to know that the meat inspection bill then forced upon Congress by a clamoring public was desired by the packers themselves. Because

Congress would not listen to the packers, and the Department of Agriculture, the Chief Executive very kindly indulged in a little conversation with a few reporters, the results of which gave Congress the needed inspiration.

It cost the Government three million dollars to tell the people that their meats are packed in a cleanly manner. if the people want this, it is all well and good. The tax it places upon the price of meat is less than half of one percent.

A similar inspection of the killing and packing of poultry would involve a very much higher rate of taxation, because of the fact that poultry products are packed in small establishments scattered throughout the entire country.

One reason that the meat packers wanted the United States inspection, is because it puts out of business the little fellow to whom the Government cannot afford to grant inspection.[*] A few of the very largest poultry packers would like to see poultry inspection for the same reason, but with the business so thoroughly scattered as to render Government inspection so expensive as to he quite impracticable, any such bill would certainly be killed in a congressional committee.

Any practical means to bring about the cleanly handling, and to prevent the consumption of diseased poultry should certainly be encouraged. This can be done by the education of the consumer. Poultry carcasses should be marketed with head and feet attached and the entrails undrawn. By this precaution the consumer may tell whether the fowl he is buying is male or female, young or old, healthy or diseased. All cold storage poultry should

---

[*] This, of course, is exactly what has happened, though the effect varies from state to state. While the USDA has broad small-producer exemptions, some states take it upon themselves to insist upon the shockingly expensive USDA inspection anyway. The result has been to drive small producers underground and to intensify the already strong anti-establishment feeling in the alternative food community.

be frozen and should be sold to the consumer in a frozen condition.

I am not in favor of the detailed regulation of business by law, but I do believe that the legal enforcement of these last precautions would be a good thing.

Chapter 11. Quality In Eggs

Because of the readiness with which eggs spoil, the term "fresh" has become synonymous with the idea of desirable quality in eggs. As a matter of fact the actual age of an egg is quite subordinate to other factors which affect the quality.

An egg forty-eight hours old that has lain in a wheat shock during a warm July rain would probably be swarming with bacteria and be absolutely unfit for food. Another egg stored eight months in a first-class cold storage room would be perfectly wholesome.

Grading Eggs

Eggs are among the most difficult of food products to grade, because each egg must be considered separately and because the actual substance of the egg cannot be examined without destroying the egg. From external appearance, eggs can be selected for size, color, cleanliness of shell and freedom from cracks. This is the common method of grading in early spring when the eggs are uniformly of good quality.

Later in the season the egg candle is used. In the technical sense any kind of a light may be used as an egg candle. A sixteen candle-power electric lamp is the most desirable. The light is enclosed in a dark box, and the eggs are held against openings about the size of a half dollar. The candler holds the egg large end upward, and gives it a quick turn in order to view all sides, and to cause the contents to whirl within the shell. To the expert this process reveals the actual condition of the egg to an extent that the novice can hardly realize. The art of egg candling cannot be readily taught by worded description. One who wishes to learn egg candling had best go to an adept in the art, or he may begin unaided and by breaking many eggs learn the essential points.

Eggs when laid vary considerably in size, but otherwise are a very uniform product. The purpose of the egg in nature requires that this be the case, because the contents of the egg must be so proportioned as to form the chick without surplus or waste, and this demands a very constant chemical composition.

For food purposes all fresh eggs are practically equal. The tint of the yolk varies a little, being a brighter yellow when green food has been supplied the hens. Occasionally, when hens eat unusual quantities of green food, the yolk show a greenish brown tint, and appear dark to the candler. Such eggs are called "grass eggs;" they are perfectly wholesome.

An opinion exists among egg men that the white of the spring egg is of finer quality and will stand up better than summer eggs. This is true enough of commercial eggs, but the difference is chiefly, if not entirely, due to external factors that act upon the egg after it is laid.

There are some other peculiarities that may exist in eggs at the time of laying, such as a blood clot enclosed with the contents of the egg, a broken yolk or perhaps bacterial contamination. "Tape worms," so-called by egg candlers, are detached portions of the membrane lining of the egg. "Liver spots" or "meat spots" are detached folds from the walls of the oviduct. Such abnormalities are rare and not worth worrying about.

The shells of eggs vary in shape, color and firmness. These variations are more a matter of breed and individual idiosyncrasy than of care or feed.

The strength of the egg shell is important because of the loss from breakage. The distinction between weak and firm shelled eggs is not one, however, which can be readily remedied. Nothing more can be advised in this regard than to feed a ration containing plenty of mineral matter and to discard hens that lay noticeably weak shelled or irregularly shaped eggs.

Preference in the color of eggs shells is well worth catering to. As is commonly stated, Boston and surrounding towns want brown eggs, while New York and San Francisco demand white eggs. These preferences take their origin from there being large henneries in the respective localities producing the particular class of eggs. If the eggs from such farms are the best in the market and were uniformly of a particular shade, that mark of distinction, like the trade name on a popular article, would naturally become a selling point. Only the select trade considers the color in buying.

Eggs of all Mediterranean breeds are white. Those of Asiatics are brown. Those of the American breeds are usually brown, but not of so uniform a tint.

The size of eggs is chiefly controlled by the breed or by selection of layers of large eggs. In a number of experiments published by various experiment stations, slight differences in the sizes of the eggs have been noted with varying rations and environment, but this cannot be attributed to anything more specific than the general development and vigor of the fowls. Pullets, at the beginning of the laying period, lay an egg decidedly smaller than those produced at a later stage in life.

The egg size table on page 161 gives the size of representative classes of eggs. These figures must not be applied too rigidly, as the eggs of all breeds and all localities vary. They are given as approximate averages of the eggs one might reasonably expect to find in the class mentioned.

How Eggs Are Spoiled

Dirty eggs are grouped roughly in three classes:
1. Plain dirties, those to which soil or dung adheres.
2. Stained eggs, those caused by contact with damp straw or other material which discolors the shell (plain dirties when washed usually show this appearance).
3. Smeared eggs, those covered with the contents of broken eggs.

For the first two classes of dirty eggs the producer is to blame. The third class originates all along the route from the nest to consumer. The percentage of dirty eggs varies with the season and weather conditions, being noticeably increased during rainy weather. In grading, about five percent of farm-grown eggs are thrown out as dirties. These dirties are sold at a loss of at least twenty percent.

The common trade name for cracked eggs is "checks." "Blind checks" are those in which the break in the shell is not readily observable. They are detected with the aid of the candle, or by sounding, which consists of clicking the eggs together. "Dents" are checks in which the egg shell is pushed in without rupturing the membrane. "Leakers" have lost part of the contents and are not only an entire loss themselves, but produce smeared eggs.

The loss from breakage varies considerably with the amount of handling in the process of marketing. A western produce house collecting from grocers by local freight will record from four to seven percent of checks. With properly handled eggs the loss through breakage should not run over one or two percent.

Eggs in which the chick has begun to develop are spoken of as "heated" eggs. Infertile eggs cannot heat because the germ has not been fertilized and can make no growth. That such infertile eggs cannot spoil is, however, a mistaken notion, for they are subjected to all the other factors by which eggs may be spoiled.

GEOGRAPHICAL CLASSIFICATION	BREED CLASSIFICATIONS	Net Wt. Per 30 Dozen Case	Weight Ounces Per Dozen	Relative Values Per Dozen
Southern Iowa's "Two ounce eggs"	Purebred flocks of American varieties of "egg farm Leghorns."	45 lbs	24	25¢
Poorest flocks of Southern Dunghills	Games and Hamburgs.	35 lbs	19.2	20¢
Average Tennessee or Texas eggs.	Poorest strains of Leghorns.	40 lbs	21.3	22¢
Average for the United States as represented by Kansas, Minnesota and Southern Illinois.	The mixed barn yard fowl of the western farm, largely of Plymouth Rock origin.	45 lbs	28	24¢
Average size of eggs produced in Denmark.	American Brahmas and Minorcas.	48 lbs.	25.6	27¢
Selected brands of Danish eggs.	Equaled by several pens of Leghorns in the Australian laying contest.	54 lbs	28.8	30¢

Figure 10. Egg size table

The sale of eggs tested out of the incubators has been encouraged by the dissemination of the knowledge that infertile eggs are not changed by incubation. Eggs thrown out of an incubator will be shrunken and weakened, and some of them may contain dead germs and the remains of chicks that have died after starting to develop. Such eggs may be sold for what they are, but should never be mixed with other eggs or sold as fresh. When carefully candled they should be worth 10¢-12¢ a dozen.[*]

Fertile eggs, at the time of laying, cannot be told from infertile eggs, as the germ of the chick is microscopic in size. If the egg is immediately cooled and held at a temperature below 70°, the germ will not develop. At a temperature of 103°, the development of the chick proceeds most rapidly. At this temperature the development is about as follows:

Twelve hours incubation: When broken in a saucer, the germ spot, visible upon all eggs, seems somewhat enlarged. Looked at with a candle such an egg cannot be distinguished from a fresh egg.

Twenty-four hours: The germ spot mottled and about the size of a dime. This egg, if not too dark-shelled, can readily be detected with the candle, the germ spot causing the yolk to appear considerably darker than the yolk of a fresh egg. Such an egg is called a heavy egg or a floater.

Forty-eight hours: By this time the opaque white membrane, which surrounds the germ, has spread well over the top of the yolk, and the egg is quite dark or heavy before the light. Blood appears at about this period, but is difficult of detection by the candler, unless the germ dies and the blood ring sticks to the membrane of the egg.

Three days: The blood ring is the prominent feature and is as large as a nickel. The yolk behind the membrane has become watery.

Four days: The body of the chick becomes readily visible, and prominent radiating blood vessels are seen. The yolk is half covered with a water containing membrane.

---

[*] This practice is not longer legal, at least in Oregon.

These stages develop as given, occurring at a temperature of 103°. As the temperature is lowered the rate of chick development is retarded, but at any temperature above 70°, this development will proceed far enough to cause serious injury to the quality of the eggs.

For commercial use eggs may be grouped in regard to heating as follows:

1. *No heat shown.* Cannot be told at the candle from fresh eggs.
2. *Light floats.* First grade that can be separated by candling, corresponding to about twenty-four hours of incubation. These are not objectionable to the average housewife.
3. *Heavy floats.* This group has no distinction from the former, except an exaggeration of the same feature. These eggs are objectionable to the fastidious housewife, because of the appearing of the white and scummy looking allantois on the yolk.
4. *Blood rings.* Eggs in which blood has developed, extending to the period when the chick becomes visible.
5. *Chicks* visible to the candle.

The loss due to heated eggs is enormous; probably greater than that caused by any other source of loss to the egg trade. The loss varies with the season of the year, and the climate. In New England, heat loss is to be considered as in the same class as loss from dirties and checks. In Texas the egg business from the 15th of June until cool weather in the fall is practically dead. People stop eating eggs at home and shipping out of the State nets the producer such small returns by the time the loss is allowed that, at the prices offered, it hardly pays the farmer to gather the eggs. In the season of 1901 hatched chickens were commonly found in cases of market eggs throughout the trans-Mississippi region, and eggs did well to net the shippers 3¢ per dozen.

Damage to eggs by heating and consequent financial loss is inexcusable. In the first place, market eggs have no business being fertilized, but whether they are or not they should be kept in a place sufficiently cool to prevent all germ growth.

The egg shell is porous so that the developing chick may obtain air. This exposes the moist contents of the egg to the drying influence of the atmosphere. Evaporation from eggs takes place constantly. It is increased by warm temperatures, dry air and currents of air striking the egg.

When the egg is formed within the hen the contents fill the shell completely. As the egg cools the contents shrink, and the two layers of membrane separate in the large end of the egg, causing the appearance of the bubble or air cell. Evaporation of water from the egg further shrinks the contents and increases the size of the air cell. The size of the air cell is commonly taken as a guide to the age of the egg. But when we consider that with the same relative humidity on a hot July day, evaporation would take place about ten times as fast as on a frosty November morning, and that differences in humidity and air currents equally great occur between localities, we see that the age of an egg, judged by this method, means simply the extent of evaporation, and proves nothing at all about the actual age.

Even as a measure of evaporation, the size of the air cell may he deceptive, for when an egg with an air cell of considerable size is roughly handled, the air cell breaks down the side of the egg, and gives the air cell the appearance of being larger than it really is. Still rougher handling of shrunken eggs may cause the rupture of the inner membrane, allowing the air to escape into the contents of the egg. This causes a so-called watery or frothy egg. The quality is in no way injured by the mechanical mishap, but eggs so ruptured are usually discriminated against by candlers.

In this connection it might be well to speak further of the subject of "white strength," by which is meant the stiffness or viscosity of the egg white. The white of an egg is a limpid, clear liquid, but in the egg of good quality that portion immediately surrounding the yolk appears to be in a semi-solid mass. The cause of this appearance is the presence of an invisible network of fibrous material. By age and mechanical disturbance this network is gradually broken down and the liquid white separates out. Such a weak and watery white is usually associated with shrunken eggs. Them eggs will not stand up well or whip into a firm froth and are thrown in lower grades.

The weakness of the yolk membranes also increases with age, and is objectionable because the breakage of the yolk is unsightly and spoils the egg for poaching.

The shrunken egg is most abundant in the fall, when the rising prices tempt the farmer and groceryman to "hold" the eggs (delay bringing them to market because waiting will bring higher prices in spite of the drop in quality). This holding is so prevalent, in fact, that from August to December full fresh eggs are the exception rather than the rule.

While we have called attention to evaporation as the most pronounced fault of fall eggs, losses from other causes are greatly increased by the holding process.

If the eggs are held in a warm place, heat and shrinkage will case the greatest damage; if held in a cellar, rot, mold, and bad odors will cause the chief loss.

The loss due to shrunken eggs is not understood nor appreciated by those outside the trade. Such ignorance is due to the fact that the shrunken is not so repulsive as the rotten or heated egg. But the inferiority of the shrunken egg is so well appreciated by the consumer that high class dealers find it impossible to use them without ruining their trade. The result is that shrunken eggs are constantly being sent into the cheaper channels, with the result that all lower grades of eggs are more depreciated in the fall of the year than at any other time.

In the classes of spoiled eggs, of which we have thus far spoken, the proverbial rotten egg has not been considered. The term "rot" in the egg trade is used to apply to any egg absolutely unfit for food purposes. But I prefer to confine the term "rotten egg" to the egg which contains a growth of bacteria.

The normal egg when laid is germ-free. But the egg shell is not germ-proof. The pores in the egg shell proper are large enough to admit all forms of bacteria, but the membrane inside the shell is germ proof as long as it remains dry. When this membrane becomes moist so that bacteria may grow in it, these germs of decay quickly grow through it and contaminate the contents of the egg.

Heat favors the growth of bacteria in eggs and sufficient cold prevents it, but as bacteria cannot enter without moisture on the surface of the egg we can consider dampness as the cause of rotten eggs. Moisture on the shell may come from an external wetting, from the "sweating" of eggs coming out of cold storage, or by the prevention of evaporation to such an extent that the external moisture of the egg thoroughly soaks the membrane. The latter happens in damp cellars, and when eggs are covered with some impervious material.

Rotten eggs may be of different kinds, according to the species of germ that causes the decomposition. The specific kinds of egg rotting bacteria have not been worked out, but the following three groups of bacterially infected eggs are readily distinguishable in the practical work of egg candling.

1. *Black rots.* It is probable that many different species of bacteria cause this form of rotten eggs. The prominent feature is the formation of hydrogen sulfide gas, which blackens the contents of the egg, gives the characteristic rotten egg smell and sometimes causes the equally well known explosion.
2. *Sour eggs or white rots.* These eggs have a characteristic sour smell. The contents become watery, the yolk and the whites mix and the whole egg is offensive to both eye and nose.
3. *The spot rot.* In this the bacterial growth has not contaminated the whole egg, but has remained near the point of entrance. Such eggs are readily picked out with the candle, and when broken open show lumpy adhesions on the inside of the shell. These lumps are of various colors and appearances. It is probable that these spots are caused as much by mold as by bacteria, but for practical purposes the distinction is immaterial.

In practice it is impossible to separate rotten from heated eggs for the reason that in the typical nest of spoiled eggs found around the farm, both causes have been at work. Dead chicks will not necessarily cause the eggs to decay, but many such eggs do become contaminated by bacteria before they reach the candler, and hence, as a physician would say, show complications.

The loss of eggs that are actually rotten is not as great as one might imagine. Perhaps one or two percent of the country's egg crop actually rot, but the expenses of the candling necessitated, and the lowering of value of eggs that contain even a few rotten specimens are severe losses.

Moldy or musty eggs are caused by accidentally wet cases or damp cellars and ice houses. The moldy egg is most frequently a spot rot. In the musty egg proper the meat is free from foreign organisms, but has been tainted by the odor of mold growth upon the shell or packing materials.

The absorption of odors is the most baffling of all causes of bad eggs. Here the candler, so expert in other points, is usually helpless. Eggs, by storage in old musty cellars, or in rooms, with lemons, onions and cheese, may become so badly flavored as to be seriously objected to by a fancy trade, and yet there is no means of detecting the trouble without destroying the egg. Such eggs occur most frequently among the held stock of the fall season.

THE LOSS DUE TO CARELESSNESS

The egg crop of the country, more than ninety-five percent of which originates on the general farm, is subject to immense waste due to ignorant and careless handling. The great mass of eggs for sale in our large cities possess to a greater or less degree the faults we have discussed.

Some idea of the loss due to the present shiftless method of handling eggs, may be obtained by a comparison of the actual average prices received for all eggs sold in New York City, and the wholesale prices quoted by a prominent New York firm dealing in high grade goods. The contrasted price for the year 1907 are shown in figure 11.

Prices at which total goods moved		Wholesale prices for strictly fresh eggs	
January	25.8	January	42
February	24.5	February	40
March	19.3	March	32
April	16.9	April	30
May	16.6	May	31
June	15.5	June	32
July	15.6	July	35
August	17.7	August	38
September	20.7	September	40
October	21.4	October	42
November	26.0	November	45
December	27.7	December	48

Figure 11. Average prices vs. strictly fresh prices

The total values figured by multiplying these prices by the New York receipts, are as follows:

Amount actually received: $23,832,000

Values at quotations for strictly fresh: $44,730,000

No one would contend it is possible to bring the entire egg crop of the country up to the latter value, but the fact that there is a definite market for eggs of first class quality at almost double the figures for which the egg crop as a whole is actually sold, is a point very significant to the ambitious producer of high grade eggs.

REQUISITES FOR THE PRODUCTION OF HIGH GRADE EGGS

1. Hens that produce a goodly number of eggs, and at the same time an egg that is moderately large (average two ounces each). Plymouth Rocks, Wyandottes, Rhode Island Reds, Orpingtons, Leghorns, Minorcas are the varieties which will do this.
2. Good housing, regular feeding and watering, and above all clean, dry nests.
3. Daily gathering of eggs, when the temperature is above 80 degrees, gathering twice a day.
4. The confining of all broody hens as soon as discovered.
5. The rejection as doubtful of all eggs found in a nest which was not visited the previous day. (Such eggs should be used at home where each may be broken separately).
6. The placing as soon as gathered of all summer eggs in the coolest spot available. *
7. The prevention at all times of moisture in any form coming in contact with the egg's shell.
8. The selling of young cockerels before they begin to annoy the hens. Also the selling or confining of old male birds from the time hatching is over until cool weather in fall.
9. The using of cracked and dirty, as well as small eggs, at home. Such eggs if consumed when fresh are perfectly wholesome, but when marketed are discriminated against and are likely to become an entire loss.
10. Keeping eggs away from musty cellars or bad odors.
11. Keeping the egg as cool and dry as possible while en route to market.
12. The marketing of all eggs at least once per week and oftener, when facilities permit.
13. The use of strong, clean cases or cartons and good fillers.

---

* Eggs should be refrigerated immediately after collection.

Chapter 12. How Eggs Are Marketed

The methods by which the larger number of American eggs pass from the producer to consumer is as follows:

The eggs are gathered by the farmer with varying regularity and are brought, perhaps on the average of once a week, to the local village merchant.

This merchant receives weekly quotations from a number of surrounding egg dealers and at intervals of from two days to two weeks, ships to such a dealer, by local freight. The dealer buys the eggs "case count"; that is, he pays for them by the case regardless of quality. He then repacks the eggs in new cases and, with the exception of a period in the early spring, candles them.

This dealer, in turn, receives quotations from city egg houses and sells to them by wire. He usually ships in carload lots. The city receiver may also be a jobber who sells to grocers, or he may sell the car outright to a jobbing house. The jobber re-candles the eggs, sorting them into a number of grades, which are sold to various classes of trade. The last link in the chain is the housewife, who, by phone or personal call, asks for "a dozen nice fresh eggs."

This most frequently repeated story of the American egg applies particularly in the case of eggs produced west of the Mississippi and marketed in the very large cities of the East.

We will now discuss the various steps of the egg trade, pointing out the reason for the existence of the present methods and their influence upon quality and consequent value.

THE COUNTRY MERCHANT

The country merchant is the logical business link between the farmer and the outside world and usually continues to act as the farmers' buyer and seller until the commodity dealt in becomes of such importance as to demand more specialized form of marketing. Eggs, being a perishable crop continuously produced, must be marketed at frequent intervals, and trips to the general store, necessary to supply the household needs, offer the only convenient opportunity for such marketing.

The merchant buys eggs because by doing so he can control his selling trade.

The farmer trades where he sells his eggs because it is convenient to do both errands at one place and also because he wishes to avoid affronting the merchant by breaking the established custom of trading out the amount.

For these reasons the merchant knows that to buy eggs means to sell goods, and he therefore bids for eggs. His competitors, across the street and in other towns, also bid for eggs.

The effect to the merchant of lowering the price of his goods or raising the price of eggs is financially the same. In either case it is the matter of cutting the prices under the spur of competition.

Now, the articles on which the merchant make his chief profits from the farmers' trade are dry goods and notions. Such articles are not standardized, but vary in a manner quite impossible of estimation by the unsophisticated. On the other hand, eggs are quoted by the dozen, and all that run may read.

Suppose, for illustration, two merchants in the same town are each doing a business with a twenty percent profit, and are buying eggs at 10¢ and selling for 11¢, the cent advance being sufficient to pay for their labor, incidental loss, and a small profit. Now one merchant decides to play for more trade. If he marks his goods down he would gain some trade, but many people would fear his goods were cheap. But if he puts up a placard, "Eleven Cents Paid For Eggs," the farmers will throng his store and never question the quality of his goods. This move having been successful, his rival across the street quietly stocks up with a cheaper line of dry goods, and some fine morning puts out a card, "Twelve Cents For Eggs." The farm wagons this week will be hitched on the other side of the street.

The rate of business at ten percent* being insufficient to maintain two men in the town, a mutual understanding is gradually

---

* A 20% profit is halved by paying 10% more for eggs.

brought about by which the prices of goods sold are worked back to the basis of twenty percent gross profit, but the false price of eggs will serve to draw the trade from neighboring towns, and is therefore maintained.

As a matter of fact the price paid to farmers for eggs by the general stores of the Mississippi Valley is frequently 1¢-2¢ above the price at which the storekeeper sells the product. Allowing the cost of handling, we have a condition prevailing in which the merchant is handling eggs at from five to ten percent loss, and it stands to reason that he is making up the loss by adding that percent to his profits on his goods. Some of the effects of this system are:

1. The inflated prices of merchandise is an injustice to the townspeople and to farmers not selling produce, in fact it amounts to a taxation of these people for the benefit of the egg producers.
2. The inflated prices of merchant's wares work to his disadvantage in competition with mail-order or out-of-town trade.
3. The farmer who exchanges eggs at inflated prices for dry goods at inflated prices is not being paid more for his eggs (save as the tax on the townspeople contributes a little to that end), but is in the main merely swapping more dollars.
4. The use of eggs as a drawing card for trade works in favor of inferior produce, and the loss to the farmer through the lowering of prices thus caused, is much greater than his gain through the forced contributions of his neighbors.

THE HUCKSTER

The huckster or peddling wagon which gathers eggs and other produce directly from the farm, prevail east and south of a line drawn from Galveston to Chicago through Texarkana, Ark., Springfield, Mo., and St. Louis. North and west of this line the huckster is almost unknown.

The huckster wagons may be of the following types:

1. An extension of the local grocery store, trading merchandise for eggs.

2. An independent traveling peddler.
3. A cash dealer who buys his load and hauls it to the nearest city where he peddles the produce from house to house or sells it to city grocers.
4. A representative of the local produce buyer.
5. A fifth style of egg wagon does not visit the farm at all, but is a system of rural freight service run by a produce buyer for the purpose of collecting the eggs from country stores.

As far as the quality of product and advantage to the farmer is concerned, the fourth style of huckster is preferable. This style exists chiefly in Indiana and Michigan and the better settled regions of Kentucky and Tennessee. The writer found hucksters in southern Michigan working on a profit of ½¢ per dozen, while in the mountains of Tennessee he found a huckster paying 10¢ for eggs that were worth 18¢ in Chattanooga, and 23¢ in New York.

The huckster scheme of gathering eggs would seemingly be a means of obtaining good eggs because of the advantage of regular collection, but in reality it does not always work out that way. While it must be admitted that in the isolated regions of the Middle and Southern States the presence of the huckster is the only factor that makes egg selling possible, it is also true that the peddling huckster of those regions usually disregards the first principles of handling perishable products. He makes a week's trip in sun and rain with his load of produce, with the result that the quality of his summer eggs is about as low as can be found.

In the more densely populated region with a twice or thrice a week, or even daily service, the huckster egg becomes the finest farm-grown egg in the market.

The second step in the usual scheme of egg marketing is the sale of eggs collected by the small storekeeper to the produce man or shipper.

THE PRODUCE BUYER

Throughout the Mississippi Valley there are wholesale produce houses at all important railroad junctions. A typical house will ship the produce of one to three counties. These houses, once a week or oftener, send out postal card quotations. These quotations read so much per case, and are usually case count, with a reservation, however, of the privilege to reject or charge loss on goods that are utterly bad. Each grocery receives quotations from one to a dozen such houses, and perhaps also from commission firms in the nearest city. The highest of these quotations gets the shipment.

The buyer repacks the eggs and usually candles them, the strictness of the grading depending upon the intended destination. The loss in candling is generally kept account of, but is seldom charged back to the shipper. The egg man wants volume of business, and if he antagonizes a shipper by charging up his loss, the usual result will be the loss of trade. So the buyer estimates his probable loss and lowers his price enough to cover it.

By "loss off," or "rots out," is meant the subtraction of the bad eggs from the number to be paid for. Buying on a "candled" or "graded basis," usually not only means rots out, but that a variation of the price is made for two or more grades of merchantable eggs.

Much discussion prevails among the western egg buyers as to whether eggs should be bought loss off or case count. Loss off buying seems to be more desirable and just, but in practice is fraught with difficulties.

If the loss off buyer feels he is losing business, he may instruct his candler to grade more closely, which means he will pay less. Whether done with honest or dishonest intent, the buyer thus sets the price to be paid after he has the goods in his own hands, and this is an obviously difficult commercial system.

Where the buyer in one case changes the grading basis to protect himself, there are probably ten cases where the eggs really deserve the loss charged; but the tenth chance gives the seller an opportunity to nurse his loss with the belief that he has been robbed by the buyer. Such an uncertain feeling is disagreeable, and the results are that where one or two competing egg dealers buys loss off, and the other case count, the case count man will get most of the business.

The case count method being the path of least resistance, the loss off system can only succeed where there is some factor that overcomes the disinclination of a shipper to let the other man set the price. This factor may be:

1. An exceptional reputation of a particular firm for honesty and fair dealing.
2. Exceptional opportunities for selling fancy goods, enabling the loss off buyer to pay much higher rates for good stuff.
3. A condition that prevails in the South in the summer, where the losses are so heavy that the dealers will not take the risk involved in case count buying.
4. Some sort of a monopoly.

A monopoly for enforcing the loss off system of buying has been brought about in some sections of the West by agreement among egg dealers. In such cases the usual experience has been that someone would get anxious for more business, and begin quoting case count, the result being that he would get the business of the disgruntled shippers in his section.

When one buyer begins quoting case count, the remainder rapidly follow suit and case count buying is quickly re-established.

THE CITY DISTRIBUTION OF EGGS

In name, city egg dealers are usually "commission houses," but in practice the majority of large lots of eggs are now bought by telegraph and the prices definitely known before shipment.

In the larger cities eggs are dealt in by a produce board of trade. Such exchanges frequently have rules of grading and an official inspector. This gives stability to egg dealing and largely solves the problem of uncertainty as to quality, so annoying to the country buyer. In the city, even where official grading is not resorted to, personal inspection of the lot by the buyer is practical, and one may know what he is getting.

In many cases, especially in smaller cities, the receiver is the jobber and sells to the grocers. In larger cities the receiver sells to a firm who makes a business of selling them to groceries, restaurants, etc.

The jobber grades the eggs as the trade demands. In a western city this may mean two grades—good and bad; in New York, it may mean seven or eight grades, the finer of these ones being packed in sealed cartons, perhaps each egg stamped with the dealer's brand.*

The city retailer of eggs include grocers, dairies, butcher shops, soda fountains, hotels, restaurants and bakeries. The soda fountain trade and the first-class hotel are among the high bidder for strictly first-class eggs. Many such institutions in eastern cities are supplied directly from large poultry farms. The figures at which such eggs are purchased are frequently at a given premium above the market quotation, or a year round contract price for a given number of eggs per week. This premium over common farm eggs may range from one or 2¢ in western cities, to 5¢-20¢ in New York and Boston. An premium of 10¢ over the quotation for Extras, or a year-round contract price of thirty-five cents per dozen, might be considered typical of such arrangements in New York City.†

---

* Sealed cartons are no longer used, as far as I know, but at least one brand (Eggland's Best) still stamps every egg. This not only makes the eggs seem fancier, it makes it a little harder for cheapskate consumers to get away with swapping a dozen fancy eggs with a dozen cheap eggs, and paying the cheap-egg price.

† To put this in perspective, "common farm eggs" were usually Grade B quality, a grade that today's consumer refuses to buy. "Extras" were Grade A quality, which in general is acceptable to today's consumer. However, both these grades had erratic quality and occasional bad eggs due to the limitations of the candling process. "Strictly Fresh" eggs were a uniform Grade AA quality. Candling errors were not an issue because the eggs were known to be newly laid.

Some of the larger chain grocers in New York City are in the market for strictly fresh eggs and have even installed buying departments in charge of expert egg men.

The great bulk of eggs move through the channels of the small restaurant, bakery and grocery. In the small cities of the Central West the grocer handles eggs at a margin of 1¢-3¢. In the South and farther West the margin is 2¢-7¢, the retail price always being in the even nickel. In the large eastern city there exists the custom, unknown in the West, of having two or more grades of eggs for sale in the same store. All eggs offered for sale are claimed by the salesman to be "strictly fresh" or "the best," and yet these eggs may vary in April from 15¢-40¢, or in December from 30¢-75¢ per dozen. The New York grocers' profit is from two 2¢-5¢ on cheap eggs, but runs higher on high grade eggs, frequently reaching 20¢ a dozen and sometimes going as high as 40¢ for very fancy stock.

City retailing is by far the most expensive item in the marketing of eggs. Figure 12 shows an illustration of the profits of the various handlers of eggs might be.

Paid the farmer in Iowa:	$0.15
Profit of country store:	$0.00
Gross profit of shipper:	$0.00 ¾
Freight to New York:	$0.00 ½
Gross profit of receiver:	$0.00 ½
Gross profit of jobber:	$0.00 ¼
Loss from candling:	$0.01 ½
Gross profit of retailer:	$0.04 ½
Cost to consumer:	$0.25

Figure 12. Breakdown of egg money.

The cheapest grades of eggs are taken by bakeries and for cooking purposes at restaurants. When cooked with other food an egg may have its flavor so covered up that a very repulsive specimen may be used. Measures have been frequently taken by city boards of health to stop the sale of spot rots and other low-grade eggs. The great difficulty with such regulations is that they are difficult of enforcement because no line of demarcation can be drawn as in the case of adulterated or preserved products.

That embryo chicks and bacterially contaminated eggs are consumed by the million cannot be doubted, but the individual examination of each egg sold would be the only way in which the food inspectors can prevent their use. The egg from the well-kept flock whose subsequent handling has been conducted with intelligence and dispatch is the only egg whose "purity" is assured, with or without law. The encouragement of such production and such handling is the proper sphere of governmental regulations in regard to this product.

COLD STORAGE OF EGGS

The supply of eggs varies from month to month, the heavy season of production centering about April and the lightest run being in November. The cold storage men begin storing eggs in March or April and continue to store heavily until June, after which time the quality deteriorates and does not keep well in storage. This storage stock begins to move out in September and should be cleaned up by December. Great loss may result if storage eggs are held too long.

The effect of the storage business is to even up the prices for the year. The reduction of the exceedingly high winter prices is unfortunate for those who are skilled enough to produce many eggs at that season of the year, but on a whole the storage business adds to the wealth-producing powers of the hen, for it serves to increase the annual consumption of eggs and prevents eggs from becoming a drug on the market during the season of heavy production.

March and April eggs are, in spite of a long period of storage, the best quality of storage stock. This is accounted for by the fact that, owing to cooler weather and rising prices, eggs leave the farm in the best condition at this season of the year.

Because eggs are spoiled by hard freezing, they must be kept at a higher temperature than meat and butter. Temperatures of from 29° to 30° F. are used in cold storage of eggs. At such temperatures the eggs, if kept in moist air, become moldy or musty. To prevent this mustiness he air in a first class storage room is kept moderately dry. This shrinks the eggs, though much more slowly than would occur without storage.

The growth of bacteria in cold storage is practically prevented, but if bacteria are in the eggs when stored they will lie dormant and begin activity when the eggs are warmed up.

Of the cold storage egg as a whole we can say it is a wholesome food product, though somewhat inferior in flavor and strength of white to a fresh egg. The cold storage egg can be very nearly duplicated in appearance and quality by allowing eggs to stand for a week or two in a dry room. Cold storage eggs, when in case lots, can be told by the candler because of the uniform shrinkage, the presence of mold on cracked eggs, and perhaps the occasional presence of certain kinds of spot rots peculiar to storage stock, but the absolute detection of a single cold storage egg is, so far as the writer knows, impossible.

It may be further said that, with the present prevailing custom of holding eggs without refrigeration for the fall rise of price, eggs placed in cold storage in April are frequently superior to the current fall and early winter receipts. Cold storage eggs are generally sold wholesale as cold storage goods, but are retailed as "eggs." The fall eggs offered to the consumer cover every imaginable variation in quality and the poorest ones sold may be a cold storage product, or they may not be.

The Bureau of Chemistry of the United States Department of Agriculture has recently announced the finding of certain crystals in the yolks of cold storage eggs that are not present in the fresh stock. This finding of a laboratory method of detecting cold storage stock was at first taken to be a great discovery. Further investigation, however, indicates that the crystal mentioned forms as the egg ages and that the rate of formation varies with the individual eggs and probably also with the temperature, so that while crystals may indicate an aged egg, the discovery only means that the microscopist in the laboratory can now do in a half hour what any egg candler in his booth can do in ten seconds.

At the present writing (February, 1909) there has been much talk of laws against the sale of cold storage eggs as fresh. The Federal Pure Food Commission, under the general law against misbranding, have made one such prosecution. Many States have agitated such laws but little or nothing has been done. I find that the idea of such a law is quite popular, especially with poultrymen. Contrary to popular opinion, the cold storage men and larger egg dealers are not opposed to the law. The people that are hit are the small dealers and especially the city grocers. These fellows buy the eggs at wholesale storage prices and sell them at retail prices for fresh, thus making excessive profits but cutting down the amount of the sales.[*] This lessens the demand for storage stock and lowers the wholesale price. This is the reason the wholesaler and warehouse man are in favor of the law.

We may all grant that the opportunity given the small dealers to grab quick gains and in so doing hurt the trade ought to be abolished. But how are we to do it? "Have State and Federal branding of the cases as they go into or come out of storage," says one—an excellent plan, to be sure by which the grocers could buy one case of fresh eggs and a back room full of storage goods and do Elijah's flour barrel trick to perfection.[†]

Clearly government inspection and stamping of each egg is the only method that would be effective and the consideration of what this means turns the whole matter into a joke. The official inspection now maintained by the boards of trade of the larger cities may be extended and the producers, dealers and consuming public may be educated to appreciate quality in eggs, as they have been in dairy products. City and State laws may also be made which will taboo the sale of spot eggs or eggs that will float on water. Meanwhile, a great opportunity is open for the man who

---

* By driving away their more discriminating customers.
† I Kings 17 (Unlike the grocer, Elijah didn't need trickery).

has high grade eggs for sale, whether he be producer or trades-
man.

Many eggs that would not do for ordinary storage are preserved by direct freezing. These eggs are broken and carefully
sorted and placed in large cans and then frozen. Such a product is
disposed of to bakers, confectioners and others desiring eggs in
large quantities. Another method of preserving eggs is by evaporation. Evaporated or dried egg is, weight considered, about the
most nourishing food product known. The chief value of such an
article lies in provisioning inaccessible regions. There is no reason,
however, why this product should not become a common article
of diet during the season of high prices of eggs. Dried eggs can be
eaten as custards, omelets, or similar dishes.*

PRESERVING EGGS OUT OF COLD STORAGE

Occasional articles have been printed in agricultural papers calling
attention to the fact that the cold storage men were reaping vast
profits which rightfully belonged to the farmer. Such writers
advise the farmer to send his own eggs to the storage house or to
preserve them by other means.

As a matter of fact the business of storing eggs has not of late
years been particularly profitable, there being severe losses during
several seasons. Even were the profit of egg storing many times
greater than they are the above advice would still be unwise, for
the storing, removing and selling of a small quantity of eggs would
eat up all possible profit.

---

* A final note on cold storage: Long-term storage isn't practiced anymore. The eggs generally come out of cold storage as Grade B quality, and consumers won't buy Grade B
eggs. Also, year-round egg production has improved tre-
mendously since 1909, so the incentive to store spring
eggs until fall has gone away. However, the freezing of liquid egg product is still practiced. In the mainstream egg
industry, liquid egg product is the dumping ground of all
eggs that don't meet Grade A standards.

The only reliable methods of preserving eggs outside of cold storage are as follows:

Liming: Make a saturated solution of lime, to which salt may be added, let it settle, dip off the clear liquid, put the eggs in while fresh, keep them submerged in the liquid and keep the liquid as cold as the available location will permit.

Water glass: This is exactly the same as liming except that the solution used is made by mixing ten percent of liquid water glass or sodium silicate with water.[*]

Liming eggs was formerly more popular than it is today. There are still two large liming plants in this country and several in Canada. In Europe both lime and water glass are used on a more extensive scale.

All limed or water glassed eggs can be told at a glance by an experienced candler. They pop open when boiled. When properly preserved they are as well or better flavored than storage stock, but the farmer or poultryman will make frequent mistakes and thus throw lots of positively bad eggs on the market. These eggs must be sold at a low price themselves, and by their presence cast suspicion on all eggs, thus tending to suppress the price paid to the producers. The farmers' efforts to preserve eggs has in this way acted as a boomerang, and have in the long run caused more loss than gain to the producers.

For the poultryman with his own special outlet for high grade goods, the use of pickling or cold storage is generally not to be considered for fear of hurting his trade. Any scheme that would help to overcome the difficulty of getting sufficient fresh eggs to supply such customers in the season of scarcity would be of great advantage. The proposition of pickling a limited number of eggs and selling them for "cooking purposes," explaining just what they

---

[*] Water glass still raises a lot of interest among people with small flocks, but apparently the quality of the eggs is quite poor, Grade B or worse, with many yolks stuck to the shell. Water glass is inferior to storage in an ordinary refrigerator.

are, ought to offer something of a solution, although, to the writer's knowledge, it has not been done.

Improved Methods of Marketing Farm-Grown Eggs

The loss to the farmers of this country from the careless handling of eggs is something enormous. No great or sudden change in this state of affairs can be brought about, but a few points on how this loss may be averted will not be out of order.

Numerous efforts have recently been made in western states to prevent the sale of bad eggs by law. Minnesota began this work by arresting several farmers and dealers. The parties invariably pleaded guilty. A number of other States followed the example of Minnesota in challenging the sale of rotten eggs, but few prosecutions were made.

Such laws mean well enough, but the only efficient means of enforcing them would be to have food inspectors who are trained as practical candlers.

The present usefulness of the laws is in calling the attention of the farmer to the mistake that he may be carelessly committing, and in placing over him a fear of possible disgrace in case of arrest and prosecution.

The weakness of the law is the difficulty of its enforcement because of the number of violations, and the difficulty of drawing distinct lines in regard to which eggs are to be considered unlawful.

Education of the farmer as to the situation is, of course, the surest means of preventing the loss, but the education of ten millions of farmers is easier to suggest than to execute. The most effective plan of education would be the introduction of a method of buying eggs similar to the one in vogue in Denmark, in which every producer is paid strictly in accordance with the quality of his eggs.

With our complicated system involving five to six dealers between the producer and the consumer, such a system is well nigh impossible. With the introduction of cooperative buying or the community system of production, paying for quality becomes entirely possible.

For enterprising farming communities, the following plans offer a cure for the evil of general store buying that take good and bad alike and causes the worthy farmer to suffer for the carelessness and dishonesty of his neighbor.

1. The encouragement of the cash buying of produce, and, if possible, the candling of all eggs with proper deduction for loss.

2. The buying of eggs by cooperative creameries. The greatest difficulty in this has been the opposition of the merchants, who through numerous ways available in a small town, may retaliate and injure the creamery patronage to an extent greater than the newly installed egg business will repay.

3. The agreement of the merchants to turn all egg buying over to a single produce buyer. This has been successfully done in a few instances, but there are not many towns in which those interested will stick to such an agreement. The worst fault with this plan is that the moment the egg buyer is given a monopoly he is tempted to lower the farmer's prices for the purpose of increasing his own profits.

4. A modification of the above scheme is the case in which the produce buyer is on a salary and in the employment of the merchants. This scheme has been successfully carried into effect in some Nebraska towns. It may be the ultimate solution of the egg buying in the West. It eliminates the temptation of the buyer to use his privilege of monopoly to fatten his own pocket-book. The weakness of the plan is that a salaried man's efficiency in the close bargaining necessary to sell the goods is inferior to that of the man trading for himself. Other difficulties are: Getting a group of merchants who will live up to such an agreement; the farmers object to driving to two places; the competition of other towns; the merchants' realization that, the farmer with cash in his pocket or a check good at all stores, is not as certain a trader as one standing, egg basket on arm, before the counter; and last, and most convincing, the merchant's further realization that any fine Saturday morning,

with eggs selling at 15¢ at the produce house, he may stick out a card saying "Sixteen Cents Paid for Eggs" and make more money in one day than his competitors did all week.

5. Cooperative egg buying by the farmers themselves. This has been discussed in a previous chapter. It is all right in localities where the business is big enough to warrant it and the farmers are intelligent and enthusiastic to back it up and stick to it.[*]

THE HIGH GRADE EGG BUSINESS

There are many excellent opportunities for men of moderate capital and ability in the high grade egg trade. The produce business on its present line, either at the country end or at the city end, is as open as any well-known form of business enterprise can be. The chances of success for a man new to the trade will be better, however, if he can find a niche in the business where he may crowd in and establish himself before the old firms realize what is up. The proposition of buying high grade eggs from producers and selling direct to consumers is a proposition of this kind.

The little game of existence is chiefly one of aping our betters and strutting before the lesser members of the flock. The large cit-

---

[*] I have leafed through *Pacific Poultryman*, the journal of the Pacific Poultry Cooperative, from between about 1925 and 1963. The cooperative ran along the lines suggested by Hastings, with every member's egg candled and graded separately and payment made according to grade. Originally, the co-op required that members sell to them exclusively. The maneuvering of other egg buyers to capture the members' business duly occurred, just as predicted by Hastings, but the cooperative quickly decided to allow its members to sell both to the co-op and to other buyers. The result was that members gleefully dumped all their worst eggs into channels buying case count (where low quality wasn't penalized), while sending all their high-quality eggs to the co-op, where high quality was rewarded. The problems caused by fast-talking outside salesmen immediately vanished.

ies are full of people in search of some way to display their superior wealth, taste and exclusiveness. If an ingenious dealer takes a dozen eggs from common candled stock, places them in a blue-lined box and labels them "Exquisite Ovarian Deposital," he can sell quite a few of them at a long price, but the game has its limits. Now, let this man secure a truly high grade article from reliable producers, teach his customers the points that actually distinguish his eggs from common stock, and he can get not only the sucker trade above referred to, but a more satisfactory and permanent trade from that class of people who are willing to pay for genuine superiority, but whose ears have not quite grown through their hats.

An express messenger running out of St. Louis became interested in the egg trade. He arranged with a few country friends to ship him their eggs. These he candled in his house cellar and began selling them to a limited trade in the wealthy section of the city. At first he delivered the eggs himself. This was in the World's Fair year of 1904. In 1908 he did a $100,000 worth of business and his type of business shows a much better percentage of profit than that of the ordinary type of dealer.

In Chicago, one of the large dairy companies established an egg department and placed a young man in charge of it. The eggs in this case are not bought of farmers but are acquired from country produce buyers whom the Chicago company have encouraged to educate their farmers to bring in a high grade of goods. These people buy their eggs in Tennessee in the winter and in Minnesota in the summer, thus getting the best eggs the year round. They sell by wagon on regular routes. The business is growing nicely and pays good profits.

Other similar concerns are operating in Chicago and other large cities. They are not numerous, however, and there is room for more. The reason the business has not been overdone is chiefly because of the difficulty of getting sufficiently really high grade eggs in the season of scarcity. Southern winter eggs are destined to relieve this situation more and more.

Another great difficulty with a plan that attempts to buy eggs directly from the producer is that premium offered on the goods tempts the farmer to go out and buy up eggs from his neighbors. This brings disastrous results in the quality of the goods and the farmer must be dropped from the list. In order to make a success, a system of buying directly from producers must be based upon a grading scheme that will pay for the actual quality of the eggs. No fear then need be exercised as to whether the farmer sells his own eggs or those of his neighbor.

The following extract from Farmer's Bulletin 128 of the U.S. Department of Agriculture has been used as advertising "dope" in the sale of high grade eggs:

"Under certain conditions eggs may be the cause of illness by communicating some bacterial disease or some parasite. It is possible for an egg to become infected with micro-organisms, either before it is laid or after. The shell is porous, and offers no greater resistance to micro-organisms which cause disease than it does to those which cause the egg to spoil or rot. When the infected egg is eaten raw the microorganisms, if present, are communicated to man and may cause disease. If an egg remains in a dirty nest, defiled with the micro-organisms which cause typhoid fever, carried there on the hen's feet or feathers, it is not strange if some of these bacteria occasionally penetrate the shell and the egg thus becomes a possible source of infection. Perhaps one of the most common troubles due to bacterial infection of eggs is the more or less serious illness sometimes caused by eating those which are 'stale.' This often resembles ptomaine poisoning, which is caused, not by micro-organisms themselves, but by the poisonous products which they elaborate from materials on which they grow.

"In view of this possibility, it is best to keep eggs as clean as possible and thus endeavor to prevent infection. Clean poultry-houses, poultry-runs and nests are important, and eggs should always be stored and marketed under sanitary conditions. The subject of handling food in a cleanly manner is given entirely too little attention."

The reprint upon the next page will give some idea of the advertising literature used in selling high grade eggs. This is a copy of a hand-bill inserted in the egg boxes of a prominent Chicago dealer:

MOORE'S BREAKFAST EGGS

are guaranteed to be perfect in quality when you receive them and to remain so until all eaten up. If for any reason they are not satisfactory return the Eggs to your dealer and get your money back.
(Signature.)

WE URGE YOU

to assist us in our endeavor to furnish you at all times with the finest Eggs by being careful to

KEEP THEM DRY.

A damp "filler" will in 24 hours make the finest fresh Eggs taste like old Cold Storage Eggs.

The flavor of an Egg cannot be detected even by the powerful electric lights used to inspect every Egg in this package, so it might be possible for a "strong" Egg to get by our inspectors, but in the past the cause of nearly every complaint has been traced to the consumer's ice box or pantry window sill.

REMEMBER

Eggs are 25¢-40¢ per doz. retail only when fine eggs are scarce. Ordinarily we can get a sufficient supply from the farmers bringing milk daily to the creameries where we make Delicia Pure Cream Butter, but in times of scarcity we often have to go as far as Oklahoma, Arkansas or Tennessee to find the best Eggs. These are not equal to our creamery Eggs but are the freshest and best to be had and are vastly superior to the old Cold Storage Eggs that flood the market at such times.

Be Sure This Seal is Unbroken When You Get the Eggs!

W. S. MOORE & CO.,
CHICAGO OFFICE—131 SOUTH WATER STREET.

BUYING EGGS BY WEIGHT

Whenever an improved method of buying is installed, eggs should be bought of the producer by weight. As far as selling to the consumer is concerned, the present scheme is more feasible; this scheme is to grade according to the size and other qualities, and sell by the dozen, the price per dozen varying according to the grade.

Buying by weight simplifies the problem of grading. It will, in addition, only be necessary to have a fine of so much for eggs that are wrong in quality. For rotten or heated eggs should be deducted an amount considerably in excess of their value, for their presence is a source of danger to the reputation of the brand. Shrunken eggs are hard to classify. In order that this may be done fairly and uniformly the specific gravity or brine test should be used. All eggs that float in a given salt brine of, say, 1.05 specific gravity should be fined. Two or more grades can be made in this fashion if desired.

THE RETAILING OF EGGS BY THE PRODUCER

In poultry papers the poultryman has been commonly advised to get near a large city and retail his own eggs at a fancy price. This sounds all right on paper but in practice it works out differently. A man cannot be in two places or do two things at the same time. The poultryman's time is valuable on his plant, and the question is whether he can handle city sales as well as a man who made it his business. If the poultryman tries to retail his own goods he will be working on too small a scale to advertise his goods or to make deliveries economically. The man making a specialty of the city end can sell ten to a hundred times as much produce as one poultryman can produce.

With a group of poultry farmers working cooperatively, or a large corporation having contracts with producers, the producing and selling end can be brought under the same management advantageously. The isolated poultryman, unless he find a market at his very door, will do better to permit at least one middleman to slip in between himself and the consumer. But there is no reason why he should not know this middleman personally and insist upon a method of buying that will pay him upon the merits of his goods.

Consigning eggs or any other produce to commission men, without a definite understanding, will always be, as it always has been, a source of dissatisfaction and loss. There is a great opportunity here for the man who can organize a system that shall do away with commission houses, other intermediate steps, and form the single step from producer to consumer. Some people say that farmers cannot be dealt with in this manner. Such people would probably have said as much about general merchandising before the days of the mail-order houses.

It is all a matter of efficient organization. A system of business fitted to deal in carload lots will, of course, fail when dealing with half cases. It is more difficult to deal in little things than in big ones because the margin is closer, but it can and will be done.[*]

THE PRICE OF EGGS

We will consider the price of all eggs from the quotation of Western firsts in the New York market. The reason for this is evident. Every egg raised east of Colorado is in line for shipment to New York. If other towns get eggs they must pay sufficiently to keep them from going to New York.

---

[*] As with selling meat, direct marketing of eggs is in vogue right now largely because few alternatives exist. Direct marketing allows you to capture the retailer's mark-up in addition to the farmer's, plus the middleman's mark-up— though middlemen generally don't have much of a mark-up. My experience is that marketing free-range eggs takes about as much time as producing them, which means that, by turning the marketing over to someone else, I could cut my margins in half, keep twice as many hens, and come out even. Using distributors, to handle one's eggs is a distinct possibility. I know one farmer who had no difficulty disposing of the output of 1,000 hens through a produce distributor. This cost him very little and eliminated about 400 miles of driving per week.

In pricing eggs we have first to consider the price of Western firsts in New York and secondly the quality relation of the particular grade to Western firsts and the consequent relation in price.

The price of eggs varies with the price of other commodities as the periods of prosperity and adversity follow one another through the years.

As is well known, all prices in the 1890's passed through a period of depression. For eggs this reached a base in 1897. Since then there has been a gradual climb till it reached a high point in 1904, remaining high until 1907. In the spring of 1908 egg prices dropped again, but the fall prices of 1908 were exceptionally high. As this work goes to press (May, 1909) eggs are going into storage at the highest May price on record.

The prices of eggs also vary independently of other commodities because of gradually changing relationships between production and consumption. As stated in the first chapter the prices of poultry products have shown a general rise when compared with other articles. This has been most marked since 1900. As for the future, we cannot prophesy except to say that there is nothing in sight to lead us to believe that we will not go still higher in egg prices.

A third variation in the price of eggs is the one caused by the seasonal relation of production and consumption. This change is from year to year fairly constant. Its normal may be seen in the scientifically smoothed curve in figure 13. This curve is based upon the New York prices for the last eighteen years.

In addition to these broader influences there are disturbing tendencies that cause the market to fluctuate back and forth across the line where the more general influences would place it.

Of those general factors, weather is the most important. Storms, rain and cold in the egg producing region decrease the rate of lay, lowering supplies and raising the price. This is due both to the fact that laying is cut down and that the country roads become impassable and the farmers do not bring the eggs to town. As long as there are storage eggs in the warehouses, weather conditions are not so effective, but when these are gone, which is usually about the first of the year, the egg market becomes highly sensitive to all weather changes. Suppose late in February storms and snows force up the price of eggs. This is followed by a warm spell which starts the March lay. The roads, meanwhile, are in a quagmire from melting snows. When they do dry up eggs come to town by the wagon load. A drop of ten cents or more may occur on such occasions within a day or two's time. This is known as the spring drop and for one to get caught with eggs on hand means heavy losses.

When once eggs have suffered this drop to the spring level, or the storage price for the season, the prices for April, May and June will remain fairly steady. About the last week in June the summer climb begins. This goes on very steadily with local variation of about the same as those of the spring months. The storage eggs begin to come out in August and at first sell about the same as fresh. As the season advances the fresh product continues to rise in price. The storage egg price will remain fairly uniform. By November the season of high prices is reached. If storage eggs are still plentiful and the weather is mild sudden variations in price may occur. These are caused by a fear that the storage eggs will not all be consumed before spring. If an oversupply of eggs has been stored a warm spell in winter will make a heavy drop in the market, but if storage eggs are scarce the sudden variations will be up-shots due to cold waves. From November until spring egg prices are a creature of the weather maps, and sudden jumps from 5¢-10¢ may occur at any time.

The price curve of 1908, which is represented by the dotted line in figure 13 will illustrate these general principles. In the lower portion of figure 13 is given the curves for the New York receipts. The heavy line represents the smoothed or normal curve, deduced from eighteen years' statistics and calculated for the year 1908. The dotted line shows the actual receipts of 1908. A comparison week by week of the receipts and price will show the detailed workings of the law of supply and demand.

Aside from the weather there are other factors that perceptibly affect the receipts and price of eggs. A high price of meat will increase farm and village consumption of eggs and cut down the receipts that reach the city. Abundance of fruit in the city mar-

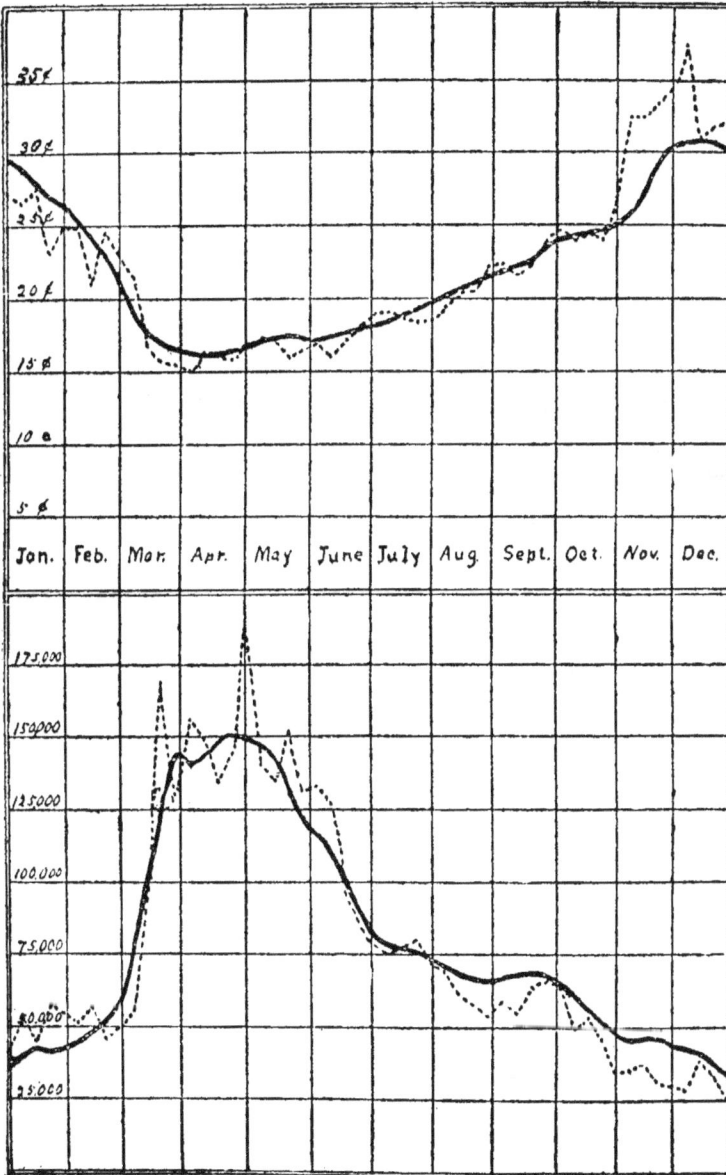

Figure 13. Egg prices and egg volume throughout the year

ket will cut down the demand for eggs. A cold, wet spring will increase the mortality of chicks and cause a decreased egg yield the following season, due to a scarcity of pullets. Scarcity and high price of feed will cut down the egg yield. High price of hens is said by some to cut down the egg yield, but I think this is doubtful, as the impulse to sell off the hens is counteracted by the desire to "keep 'em and raise more."

Figure 14 gives the quotations taken from the *New York Price-Current* for November 14, 1908.

The writer was in the New York market at the time and saw many cases of White Leghorn eggs sell wholesale at as high as 55¢. These were commonly retailed at 5¢ each (60¢ per dozen). There were a good many brands retailing at 65¢ and one of the largest high class groceries was selling for 70¢. This is practically double the official quotations and three times that of cold storage stock.

The above prices represent a fair sample of the fall prices of 1908. It should be noted that the 1908 fall prices were relatively somewhat better than the rest of the season.

The time of high prices is also the time of the greatest variation in the price of the different grades. In the springtime all eggs are fairly fresh and good, and the fanciest eggs bring wholesale only 2¢ or 3¢ above quotations. There are a few retailers who hold the spring prices to their customers up above the general market. One New York firm that does a large high class egg business never lets their price at any season go below 40¢. This, of course, means big profits and sales only to those who, when they are satisfied, never bother about price.

In the fall any man who has fresh eggs can sell them at very near the highest price, but in the spring only a small percent can go at fancy prices and the great majority of even the high grade eggs must go at very ordinary prices. In the summer months there is not so much demand in the cities, as the wealthy are not there to buy. The coast and mountain resorts are then good markets for fancy produce.

N. Y. Mercantile Exchange Official Quotations

State, Pennsylvania and nearby fresh eggs continue in very small supply and of more or less irregular quality, a good many being mixed with held eggs—sometimes with pickled stock. The few new-laid lots received direct from henneries command extreme prices—sometimes working out in a small way above any figures that could fairly be quoted as a wholesale value. We Quote: Selected white, fancy, 48-50¢/doz.; fair to choice, 35-46¢./doz.; lower grades, 26-32¢/doz.; brown and mixed, fancy, 38-40¢/doz.; fair to choice, 30-36¢/doz.; lower grades, 25-28¢/doz.

Type	Price, ¢
Fresh gathered, extras, per dozen	37
Fresh gathered, firsts	32-33
Fresh gathered, seconds	29-31
Fresh gathered, thirds	25-28
Dirties, No. 1	21-22
Dirties, No. 2	18-20
Dirties, inferior	12-17
Checks, fresh gathered, fair to prime	18-20
Checks, inferior	12-16
Refrigerator, firsts, charges paid for season	24-24½
Refrigerator, firsts, on dock	23-23½
Refrigerator, seconds, charges paid for season	22½-23½
Refrigerator, seconds, on dock	21½-2½
Refrigerator, thirds	20-21
Limed, firsts	22½-23
Limed, seconds	21-22

Figure 14. Price quotes for November 14, 1908

Notes On Chapter 12

At Norton Creek Farm, we've found that we don't have much competition from other high-grade egg producers, since there aren't many of them and all of them have much smaller flocks than we do. But our seasonal variation in egg production means that we have too many eggs in the spring and too few in the fall. This used to be made worse by our desire to have steady prices year-round.

In 2002, we finally realized that maintaining steady prices was harmful to both our customers and ourselves. Now, if egg begin to pile up, we lower prices, and if the shelves are empty when we make deliveries, we raise prices. This is about the simplest example of supply and demand there is.

This method works very well. If we kept our prices high in the spring, a lot of our eggs would remain unsold. Dropping the price causes them to sell briskly. If we did not raise prices in the fall and winter, the shelves would be empty most of the time, frustrating customers who would willingly pay more, if that's what it takes to get what they want. By raising prices, at least the customers have a choice; to buy or not to buy. An empty shelf robs them of choice.

New farmers come in sometimes and undercut us. Normally they don't have enough volume to affect our sales very much, though consumers tend to be bargain hunters can often be led astray for a nickel. But producing first-quality free-range eggs is labor-intensive, and farmers who offer them at low prices generally have to work awfully hard for their money, and they tend not to increase their flocks or even replace their old hens when the time comes. The low return on their labor robs them of incentive.

Because consumers love low prices and sales, these are a good way of attracting new customers. The nice thing about a sale is that customers see it as a promotional gimmick rather than a permanent drop in prices, and you can have frequent sales or rotating specials (Large eggs one week, Jumbos the next) without lowering their idea of the "fair price" of your eggs very much.

In any event, the best way to justify high prices is to have better eggs than anyone else. At the very least, they should be cleaner and colder than any other local producer's.

Chapter 13. Breeds of Chickens

I do not place much dependence on the results of breed tests. Indeed, I consider the almost universal use of the Barred Rock in the most productive farm poultry regions in the United States, and the equal predominance of Single-Comb White Leghorns on the egg farms of New York and California, as far more conclusive than any possible breed tests.

BREED TESTS

In Australia there has been conducted a series of breed tests so remarkable and extensive that the writer considers them well worth quoting. The Hawkesberry Agricultural College Tests extend over a period of five years, the pens entered were of six birds each, and the time one year. The results were as follows:

Year	No. of Pens Competing	Yield of Highest Pen	Average Yield of All Pens
1903	70	218	163
1904	100	204	152
1905	100	235	162
1906	100	247	177
1907	60	245	173

The winners and losers for five years were as follows:

Year	Winning Pen	Losing Pen
1903	Silver Wyandotte	Silver Wyandotte
1904	Silver Wyandotte	Partridge Wyandotte
1905	Single Comb White Leghorns	Single Comb White Leghorns
1906	Black Langshans	Golden Wyandotte
1907	Single Comb White Leghorns	Single Comb Black Leghorns

As a matter of fact, the winning pen means little for breed comparison. This is shown by the winning and losing pens frequently being of the same breed.

The average for hens of one breed for the whole five years is more enlightening. For the three most popular Australian breeds, these grand averages are:

Breed	No. Hens	Average Egg Yield	Av. Wt. Eggs Oz. per Doz.
Single Comb White Leghorns	564	175.5	26.4
Black Orpingtons	522	166.6	26.1
Silver Wyandottes	474	161.1	24.9

These figures are undoubtedly the most trustworthy breed comparisons that have ever been obtained. When we go into the other breeds, however, with smaller numbers entered, the results show chance variation and become untrustworthy, for illustration: Rose Comb Brown Leghorns, with 42 birds entered, have an average of 176.4. This does not signify that the Rose Comb Browns are better than the Single Comb Whites, for if the Whites were divided by chance into a dozen lots of similar size, some would undoubtedly have surpassed the Rose Comb Browns. As further proof, take the case of the Rose Comb White Leghorns with 36 birds entered and an egg yield of 166.9. Both breeds are probably a little poorer layers than Single Comb Whites, but luck was with the Rose Comb Browns and against the Rose Comb Whites. For a discussion of this principle of the worth of averages from different sized flocks see Chapter XV.

All Leghorns in the tests, with 846 birds entered, averaged 170.3 eggs each. All of the general-purpose breeds (Rocks, Wyandottes, Reds and Orpingtons), with 1416 birds entered, averaged 160.2. The comparison between the Leghorns and the general-purpose fowls as classes is undoubtedly a fair one. A study of the relations between the leading breeds in these groups and the general average of these groups is worthwhile.

It bears out the writer's statement that the best fowls of a group or breed are to be found in the popular variety of that breed. The Australian poultryman, wanting utility only, would do well to choose out of the three great Australian breeds here mentioned. The Single Comb White Leghorn is the only one of the three breeds to which the advice would apply in America. Barred Rocks and perhaps White Wyandottes, would here represent the other types.

There is one more point in the Australian records worthy of especial mention. The winning pen in 1906 were Black Langshans and, what seems still more remarkable, were daughters of birds purchased from the original home of Langshans in North China. Other pens of Langshans in the test failed to make remarkable records, but this pen of Chinese stock, with a record of 247 eggs per hen for the first year and 415 eggs per hen for two years, is the world's record layers beyond all quibble. This record is held by a breed and a region in which we would not expect to find great layers.

This holding of the record by a breed hitherto not considered a laying type, would be comparable to a tenderfoot bagging the pots in an Arizona gambling den. If the latter incident should occur and be heralded in the papers it would be no proof that it would pay another Eastern youth to rush out to Arizona. It is probable that the man who, on the strength of this single record, stocks an egg farm with imported Chinese Langshans, will fare as the second tenderfoot.

The year following the Langshan winning, the first eleven winning pens were all Single Comb White Leghorns. This is also remarkable—much more remarkable in fact than the Langshans record. It is like a royal flush in a poker game. Standing alone, this would be very suggestive evidence of the eminence of the breed. Standing as it does, with the combined evidence of years and numbers, it gives the Single Comb White Leghorn hen the same reputation in Australia as she has in America and Denmark—that of being the greatest egg machine ever created.

Isolated evidence is misleading. Accumulated evidence is convincing. The difference between the scientist and the enthusiast is that the former knows the difference between these two classes of evidence.

THE HEN'S ANCESTORS

To one who is unfamiliar with the different types of chickens found in a poultry showroom, it seems incredible that those varieties should have descended from one parent source. It was, however, held by Darwin that all domestic chickens were sprung from a single species of Indian jungle fowl. Other scientists have since disputed Darwin's conclusion, but it does not seem to the writer that the origin of domestic fowls from more than one wild variety makes the changes that have taken place under domestication any less remarkable.

The buff, white and dominique colors, unheard of in wild species, frizzles with their feathers all awry, the Polish with their deformed skulls and the sooty fowls whose skin and bones are black, are some of the remarkable characters that have sprung up and been preserved under domestication. The varieties of domestic fowl form one of the most profound exhibits of man's control over the laws of inheritance. What makes these wonders all the more inexplicable is that these profound changes were accomplished in an age when a scientific study of breeding was a thing unheard of.

The wild chicken whom Darwin credits as the parent of the modern gallinaceous menagerie, is smaller than modern fowls and is colored in a manner similar to the Black-breasted Game. The habits of this bird are like those of the quail and prairie-chicken, both of which belong to the same zoological family.

From its natural home in India the chicken spread east and west. Chinese poultry culture is ancient. In China, as well as in India, the chief care seems to have been to breed very large fowls, and from these countries all the large, heavily feathered and feather legged chickens of the modern world have come.

Poultry is also known to have been bred in the early Babylonian and Egyptian periods. Here, however, the progress was in a different line from that of China. Artificial incubation was early developed, and the selection was for birds that produced eggs continually, rather than for those that laid fewer eggs and brooded in the natural manner.

The Egyptian type of chicken spread to the countries bordering on the Mediterranean, and from Southern Europe our non-sitting breeds of fowls have been imported. Throughout the countries of Northern Europe minor differences were developed. The French chickens were selected for the quality of the meat, while in Poland the peculiar top-knotted breed is supposed to have been formed.

The English Dorking is one of the oldest of European breeds and is possessed of five toes. Five-toed fowls were reported in Rome and exist today in Turkey and Japan. The Dorkings may be descended directly from the Roman fowls, or various strains of five-toed fowls may have arisen independently from the preservation of sports.

The chief point to be noted in all European poultry is that it differs from Asiatic poultry in being smaller, lighter feathered, quicker maturing, of greater egg-producing capacity, less disposed to become broody, and more active than the Asiatic fowl.

The early American hens were of European origin, but of no fixed breeds. About 1840 Italian chickens began to be imported. These, with stock from Spain, have been bred for fixed types of form and color, and constitute our Mediterranean or non-sitting breeds of the present day. Soon after the importation of Italian chickens a chance importation was made from Southeastern Asia. These Asiatic chickens were quite different from anything yet seen, and further importations followed.

Poultry-breeding soon became the fashion. The first poultry show was held in Boston in the early 1850s. The Asiatic fowls imported were gray or yellowish-red in color, and were variously known as the Brahmapootras, Cochin-Chinas and Shanghais. With the rapid development of poultry-breeding there came a desire to produce new varieties. Every conceivable form of cross-breeding was resorted to. The great majority of breeds and varieties as they exist today are the results of crosses followed by a few years of selection for the desired form and color. Many of our common breeds still give us occasional individuals that resemble some of the types from which the breed was formed. The exact history of the formation of the American or mixed breeds is in dispute, but it is certain that they have been formed from a complex mixing of blood from both European and Asiatic sources.

The English have recently furnished the world with a very popular breed which was originated by the same methods. I refer to the Orpingtons.

The ever-growing multiplicity of varieties of chicken is in reality only casually related to the business of the poultryman whose object is the production of human food.

Breeding as an art or vocation is a source of endless pleasure, and, as such, is as worthy of encouragement as is painting, music, or the collection of the bones of prehistoric animals. Breeding as an art has produced many forms of chickens that are entirely worthless as food producers, but this same group of poultry breeders, tempered to be sure by the demands of commercialism, has produced other breeds that are certainly superior for the various commercial purposes to the unselected fowls of the old-fashioned farm-yard.

The mongrel chicken is a production of chance. Its ancestry represents everything available in the barn-yard of the neighborhood, and its offspring will be equally varied. In the pure breeds there has been a rigid selection practiced that gives uniform appearance. The size and shape requirements of the standard, although not based on the market demands, come much nearer producing an ideal carcass than does chance breeding. Ability to mature for the fall shows is a decidedly practical quality that the fancier breeds into his chickens. Moreover, poultry-breeders, while still keeping standard points in mind, have also made improvements in the laying and meat-producing qualities of their chickens. Considering these facts it is an erroneous idea to think that mongrel chickens offer any advantage over pure-bred stock.

In the broader sense we may regard as pure-bred those animals that reproduce their shape, color, habits, or other distinctive qualities with uniformity. In order that we may get offsprings like the parent and like each other, we must have animals whose ancestors for many generations back have been of one type. The more generations of such uniformity, the more certain it will be that the young will possess similar quality.

One strain of chickens may be selected for uniform color of feathers, another for a certain size and shape, another for laying large eggs of a certain color, and yet another strain for being producers of many eggs. Each of these strains might be well-bred in these particular traits, but would be mongrels when the other considerations were taken into account.

This explains to us why the family or strain is frequently more important than the breed. In fact, the whole series of breed classification is arbitrary. This is especially true of the American or mixed breeds. Humorously turned fanciers at the poultry show frequently have much sport trying to get other fanciers to tell White or Buff Rocks from Wyandottes when the heads are hidden. From the dressed carcasses with feet and head removed, the finest set of poultry judges in the world would be hopelessly lost in a collection of Rocks, Wyandottes, Reds and Orpingtons and, I dare say, one could run in a few Langshans and Minorcas if it were not for their black pin feathers.

WHAT BREED

The writer has great admiration for breeding as an art. He would rather be the originator of a breed of green chickens with six toes than to have been the author of "Afraid to Go Home In the Dark." But I do want the novice who reads this book to be spared some of the mental throes usually indulged in over the selection of a breed.

So-called meat breeds; that is, the big feather-legged Asiatics, save on a few capon and roaster plants in New England, are really useless. They have given size to American chickens as a class, and in that have served a useful purpose, but standing alone they cannot compete with lighter, quick-growing breeds.

For commercial consideration there are really only two types: The egg breeds of Mediterranean origin and the general-purpose breeds or "growers," including the Rocks, Wyandottes and Rhode Island Reds. The difference between the layers on the one hand and the growers on the other is quite important. Which should be used depends on the location and plan of operations, as has already been discussed.

The choice of variety within the group is a matter of taste and chance of sales of fancy stock. This one principle can, however, be laid down: The more popular the breed, the more choice there will be in selecting strains and individuals. Pea Comb Plymouth Rocks and Duckwing Leghorns should not be considered because of their rarity. Of the growers, their popularity and claims are close enough to make the particular choice unimportant. For commercial consideration, the writer would as soon invest his money in a flock of Barred Rock, White Wyandottes or Rhode Island Reds. Among layers the Single Comb White Leghorn has achieved such a lead that the majority of good laying strains are in this breed and to choose any other would be to place a handicap on oneself. For a description of breeds, the reader should secure an *Illustrated American Standard of Perfection,* or some of the books published by poultry fanciers and judges. To take up the matter here would merely be using my space for imparting knowledge which can be better secured elsewhere.

The relative popularity of breeds at the poultry shows is nicely shown by the following list. This data was compiled by adding the numbers of each breed exhibited at 124 different poultry shows in the season of 1907. A detailed report of the total entries of each breed is as follows: Plymouth Rocks, 14,514; Wyandottes, 12,320; Leghorns, 8,740; Rhode Island Reds, 5,812; Orpingtons, 2,857; Langshans, 2,153; Minorcas, 1,709; Cocoon Bantams, 1,590; Games, 1,277; Brahmas, 1,181; Cochins, 1,010; Hamburgs, 758; Game Bantams, 637; Polish, 618; Houdans, 538; Indians, 538; Anconas, 464; Sebright Bantams, 423; Andalusians, 117; Japanese Bantams, 115; Dorkings, 105; Brahma Bantams, 104; Buckeyes, 95; Silkies, 85; Spanish, 83; Redcaps, 71; Sumatras, 41; Polish Bantams, 37; Sultans, 18; Malays, 12; Frizzles, 7; Le Fleche, 7; Dominiques, 5; Booted Bantams, 4; Malay Bantams, 3; Crevecoeure, 3.

Chapter 14. Practical and Scientific Breeding

Science has been defined as the "know how" and art as the "do how." The man who works by art depends upon an unconscious judgment which is inborn or is acquired by long practice. The man who works by science may also have this artistic taste, but he tests its dicta by comparison with known facts and principles. The scientist not only looks before he leaps, but measures the distance and knows exactly where he is going to land.

Breeding has for centuries been an art, but the science of breeding is so new as to seem a mass of contradictions to all except those familiar with the maze of mathematics and biology by which the barn-yard facts must find their ultimate explanation. The science of breeding may in the future bring about that which would now seem miraculous, but it is the ancient art of breeding that is and will for years continue to be the means by which the poultry fancier will achieve his results.

In a volume the chief aim of which is to place the poultry industry, which is now conducted as an art, in the realm of technical science, it might seem proper to devote considerable space to the subject of breeding. That I shall not do so, is for the reason that while theoretically I recognize the important part that breeding plays in all animal production, for the practical proposition of producing poultry products at the lowest possible cost, a knowledge of the technical science of breeding is inessential and may, by diverting the poultryman's time to unprofitable efforts, prove an actual handicap.

For the show-room breeder the new science of breeding is too undeveloped to be of immediate service, or I had better say, the show room requisites are too complicated for theoretical breeding to promise results. For the commercial poultryman, I shall review what has been accomplished and state briefly the theories upon which contemplated work is based.

The objects strived for in poultry breeding are:

1. To create new varieties which shall have improved practical points or shall attract attention as curiosities.
2. To approach the ideals accepted by fanciers for established breeds, and hence win in competition.
3. To change some particular feature or habit as, to increase egg production or reduce the size of bantams.
4. To improve several points at once as, eggs and size in general purpose fowls.

This classification is really unnecessary, as the most specialized breeding involves consideration of many points.

BREEDING AS AN ART

The method by which breeds and varieties of the show room specimens have been developed is essentially as follows: The wonderfully different varieties of fowl from every quarter of the earth are brought together. Crossing is then resorted to, with the result that birds of all forms and colors are produced. The breeder then selects specimens that most nearly conform to the type or ideal in his mind.

Suppose a man wished to produce Barred Leghorns with a fifth toe. He would secure Barred Rocks, White Leghorns and White or Gray Dorkings. Then he would cross in every conceivable fashion.

Perhaps he might have trouble getting the white color to disappear. In that case Buff Leghorns, which are a newer breed, might be tried and found more pliable material. By such methods the breeder would, in three or four generations of crossing, get a crude type of what he desired. Henceforth it would be a matter of patience and selection. Five to twenty years is the time usually taken to produce new breeds of fancy poultry that will breed true to type. In this style of breeding the principles at stake are simple. The first is to secure the variations wanted; second, to breed from the most desirable of these specimens.

The same methods of selection that establish a breed are used to maintain it, or to establish strains. In ordinary breeding there are two other principles that are sometimes called into play. One is prepotency, the other is inbreeding. By prepotent we mean having unusual power to transmit characters to offspring. Suppose a breeder has five yards headed by five cock birds. The male in yard two he does not consider quite as fine as the bird in yard one, but in the fall he finds the offspring of bird from two much better than the offspring from yard one. The breeder should keep the prepotent sire and his offspring rather than the more perfect male, who fails to stamp his traits upon his offspring.

Normally a child has two parents, four grandparents, and eight great-grandparents. Now, when cousins marry, the great-grand-parents of the offspring are reduced to six. The mating of brother and sister cuts the grandparents to two, and the great-grandparents to four. Mating of parent and offspring makes a parent and grandparent identical and likewise eliminates ancestry. Inbreeding means the reduction of the number of branches in the ancestral tree, and this means the reduction of the number of chances to get variation, be they good or bad.

Inbreeding simply intensifies whatever is there. It does not necessarily destroy the vitality, but if close inbreeding is practiced long enough, sooner or later some little existing weakness or peculiarity would become intensified and may prove fatal to the strain. For illustration, suppose we began inbreeding brother and sister with a view of keeping it up indefinitely. Now, in the original blood, a tendency for the predominance of one sex over the other undoubtedly exists and would be intensified until there would come a generation all of one sex, which, of course, terminates our experiment.

Inbreeding has always been tabooed by the people generally. Meanwhile the clever stock breeders have combined inbreeding with selection and have won the show prizes and sold the people "new blood" at fancy prices.

Unintelligent inbreeding, as practiced on many a farm, results in run-down stock, not so much from inbreeding as from lack of selection. Out-crossing or mixing in of new blood is better than hit-or-miss inbreeding. Intelligent inbreeding is better still.

SCIENTIFIC THEORIES OF BREEDING

The main tenet of Darwin's theory of racial inheritance or evolution was that changes in animal life, wild or domestic, were brought about by the addition of very slight, perhaps imperceptible, variations. He argued that the giraffe with the longest neck could browse on higher leaves in time of drought and hence left offspring with slightly longer necks than the previous generation.

Upon this theory the ordinary breeding by selection is based. In case of breeding for the show room, the breeder's eye, or the Judge's score card, is the tape with which to measure the length of the giraffe's neck. This principle can be applied equally well, even better, to characteristics where accurate measurement may be used.

The last forty years of scientific progress has established firmly the general theories of Darwin, but they have also resulted in our questioning his idea that all great changes are due to the sum of small variations. Many instances have been suggested in which the theory of gradual changes could not explain the facts.

The theory of mutation, of which Hugo de Vries of Holland is the chief expounder, does not antagonize Darwin, but simply gives more weight in the process of evolution to the factor of sudden changes commonly called "sports." Let us illustrate: in the giraffe of our former forest, one might appear whose neck was not longer because of slightly longer vertebrae, but who possessed an extra vertebra. This would be a mutation. In other words, a mutation is a marked variation that may be inherited. We now believe that polled cattle, five-toed Dorkings, top-knotted Houdans, frizzles and black skinned chickens arose through mutations.

Burbank's Methods—The wonderful Burbank, with his thornless cactus, his stoneless plum, and his white blackberry, is simply a searcher after mutations. His success is not because he uses any secret methods, but because of the size of his operations. He produces his specimens by the millions, and in these millions looks (and often looks in vain) for the lonely sport that is to father a new race. Burbank has, with plants, many advantages of which the animal breeder is deprived. He can produce his specimens in greater number, he can more easily find out the desirable character, and in many plants he has not the uncertain element of double parentage to contend with, while with others he is still more fortunate, as he can produce them by seed, stimulate variation until the desired mutation is found and can then reproduce the desired variation with certainty by the use of cuttings. This latter is not true inheritance with its inevitable variation, but the indefinite prolongation of the life of one individual. In this sense there is only one seedless orange tree in the world.

The Centgenitor System—Prof. Hays in breeding wheat at Minnesota, first used in this country a system of breeding which is essentially as follows: A large variety of individual seeds are selected. These are planted separately and the amount and character of the yield observed. The offspring of one seed is kept separate for several generations, or until the character of the tribe is thoroughly established. The advantage of this plan of breeding is in that the selection is not made by comparing individuals, but by comparing the offspring of individuals. Thus, we necessarily select the only trait really worth while; that is prepotency or the ability to beget desirable qualities.[*]

---

[*] Here Hastings is anticipating the use of progeny testing as the basic tool of breed improvement. It was almost fifty years before this method took over completely. Breeders clung to the discredited system of assuming that high-producing hens would automatically have high-producing offspring. This sounds right but doesn't work in practice. Most high-producing hens have average offspring. Many governments forced such discredited methods upon their breeders, greatly hampering their efforts. One reason the U.S. became dominant in the poultry world was that membership in the official breed improvement plan was voluntary, and the best breeders refused to participate.

The application of this centgenitor system necessitates inbreeding; it also necessitates large operations. Of the former, breeders have generally been afraid; of the latter they have lacked opportunity. But the centgenitor system, combined with Burbank's principle of large opportunity of selection, is, in the writers belief, the method by which the 200-egg hen will be ultimately established in America.[*]

Much of the recent stimulus to the study of the Science of Breeding was occasioned by the discovery of Mendel's Law. Briefly, the law states that when two pure traits or characters are crossed, one dominates in the first generation of offspring—the other remaining hidden or recessive. Of the second generation, one-half the individuals are still mixed, bearing the dominant characteristic externally and the other hidden; one-fourth are pure dominants and one-fourth are pure recessives. In future generations the mixed or hybrid individuals again give birth to mixed and pure types apportioned as before, thus continuing until all offspring become ultimately pure. For illustration: If rose and single comb chickens are crossed, rose combs are dominant. The first generation will all have rose combs. The second generation will have one-fourth single combs that will breed true, one-fourth rose combs that will breed rose combs only, and one-half that again will give all three types.

---

[*] The first 200-egg flock was created within a short drive from my farm, at Oregon Agricultural College (later Oregon State University), by Prof. James Dryden, using methods of which Hastings would no doubt have approved, including breed crosses. At the time, crosses between two breeds of chicken were considered by many to be against nature, and a bill was introduced into the Oregon State Legislature to forbid it, though without success. The three breeds developed by Dryden were so superior that proceeds from the sale of breeding stock paid for many of the buildings on the OSU campus. Dryden's book, *Poultry Breeding and Management* (1916), remained in print through the Forties, and was considered by many to be the most popular poultry book of all time.

Mendel's Law works all right in cases where pure unit characteristics are to be found. For the great practical problems in inheritance, Mental's law is utterly hopeless. The trouble is that the chief things with which we are concerned are not unit characteristics but are combinations of countless characteristics which cannot be seen or known, hence cannot be picked out. Thus the tendency to revert to pure types is foiled by the constant recrossing of these types.

Mendel's law is a scientific curiosity like the aeroplane. It may some day be more than a curiosity, but both have tremendous odds to overcome before they supplant our present methods.

Prof. C. B. Davenport, of the Carnegie Institute, is working on experimental poultry breeding in its purely scientific sense. His conclusions have been much criticized by poultry fanciers. The truth of the matter is that the fancier fails to appreciate the spirit of pure science. The scientist, enthused to find his white fowl reoccur after a generation of black ones, is wholly undisturbed by the fact that the white ones, if exhibited, might be taken for a Silver Spangled Hamburg.

Mendel's law as yet offers little to the fancier and less to the commercial poultryman. Its study is all right in its place, but its place is not on the poultry plant whose profits are to buy the baby a new dress.

BREEDING FOR EGG PRODUCTION

Attempts to improve the egg-producing qualities of the hen date from the domestication of the hen, but it has only been within the last few years that rapid progress has been possible in this work. The inability to determine the good layers has been the difficulty.

The great majority of people make no selection of hens from which to hatch their stock. The eggs of the whole flock are kept together and when eggs are desired for hatching they are selected from a general basket. It has been assumed, and is shown by trap-nest records, that eggs thus selected in the spring of the year are from the poorer rather than from the better layers. This is because hens that have not been laying during the winter will lay very heavily during the spring season. Many breeders have attempted to pick out the good layers by the appearance of the hens. Before the advent of the trap-nest the "egg type" of hen was believed to be a positive indication of a good layer. The "egg type" hen had slender neck, small head, long, deep body of a wedge shape. Various "systems" founded on these or other "signs" have been sold for fancy prices to people who were easily separated from their money. Trap-nest records show such systems to be on a par with the lunar guidance in agricultural operations.

I might remark here that the determination of sex by the shape of the egg or similar methods, is in a like category. Science finds no proof of such theories.

A few methods of selecting the layers have been suggested which, while far from absolute, are of some significance and are well worth noting. The hen that sits upon the roost while other hens are out foraging, is probably a drone. The excessively fat or the excessively lean hens are not likely to be layers. It would naturally be supposed that the active laying hen would be the last one to go to roost at night. At the Kansas Experiment Station, the writer made observations upon the order in which the hens went to roost, and the above assumption was found in the majority of cases to be correct.

A still better scheme of selecting layers is the practice of picking out the thrifty, quickly maturing pullets when they first begin to lay in the fall season. At the Maine Experiment Station, such a selection gave a flock of layers which averaged about one hundred and eighty eggs, when the remainder of the flock yielded only one hundred and forty.[*]

Trap-nests devised to catch the hen that lays the egg are numerous in the market. A trap-nest to be successful, must not only catch the hen that lays, but must prevent the entrance of the other hens. The more trap-nests that are provided, the less often they will require attention; but the more often the nests are attended, the better for the comfort of the hens.

---

[*] At the time, a flock giving 180 eggs per year was extremely productive.

The use of trap-nests is expensive and cannot be recommended for the poultryman who must make every hour of time put on his chickens yield him an immediate income. Fanciers and Experiment Stations can well afford to use trap-nests and must, indeed, use them both for breeding for egg production, and also for determining the hen that laid the egg when full pedigrees are desired in other breeding work.

A scheme that has sometimes been used in the place of trap-nests, is a system of small compartments, in each of which one hen is kept. Such a scheme does not seem feasible on a large scale, but for breeders wishing to keep the records of a small number of hens, it is all right. Because of its cost, this system is wholly out of the question, except for a man following breeding as a hobby and who cannot devote himself during the day to the care of trap-nests.

Having determined the best layers, it remains to breed from these and from their descendants. The tests of pullets hatched from hens are better signs of the hen's value as a breeder than is her own record. It has been surmised that a hen which lays heavily will not lay eggs containing vigorous germs. So far as the writer's experience has gone, the laying of infertile eggs is a family or individual trait not particularly related to the number of eggs laid.

When we have bred from the best layers and have raised our average egg yielder to a higher level, the question arises as to whether the strain will permanently maintain the high yield or drop back to the former rate of production. Theory says that it will not drop back. As a matter of fact it will do so, for the heavier production will be more trying on the hen's constitution, and naturally selection will gradually cause the egg record to dwindle. Hence the necessity of continued selection or the infusion of new blood from other selected strains. [*]

---

[*] Theory is correct here, and Hastings is wrong. The issue was clouded at the time by the lack of knowledge about vitamins. High-producing hens continue to lay in spite of obstacles such as poor winter nutrition, and this is hard on their health.

Whatever may be the change desired in a strain of chickens, specimens showing the trait to be selected should be used as breeders. Those characteristics readily visible to the eye have long been the subjects of the breeder's efforts. But traits not directly visible can likewise be changed by breeding. The number of eggs, size and color of eggs, rapid growth, ready fattening powers, quality of meat and general characteristics, are all matters of inheritance, and if proper means are taken to select the desirable individuals all such characteristics can be changed at the will of the breeder.

It is a fact, however, often overlooked, that the more traits for which one selects, the slower will be progress. For illustration: if in breeding for egg production, one-half the good layers are discarded for lack of fancy points, the progress will be just half as rapid.[*]

A discussion of the work in breeding for egg production at the Maine Experiment Station is taken up in the next chapter.

---

[*] For just this reason, utility poultry and fancy poultry parted ways gradually over the years, though slowly, for there were many holdouts. It makes sad reading to leaf through journals of the American Poultry Association in the Fifties, when the editors continually predicted that hybrid chickens and utility strains that were not bred for show were just a passing fad, and that the faithful would, in the end, be rewarded.

Chapter 15. Experiment Station Work

Our entire scheme of agricultural education and experimentation is new. The poultry work at experiment stations is very new. Ten years will about cover everything worthy of a permanent record in the poultry experiment station files.

Stations Leading in Poultry Work

Among the earliest stations to begin poultry work in this country were Rhode Island, Massachusetts, Connecticut and Maine. Rhode Island conducted the first school of poultry culture. The two stations of New York State were also early in the work, and Cornell now has the leading school of poultry culture in this country.

West Virginia has always maintained a considerable poultry plant. Outside of the states east of the Appalachians, the first poultry work to be heard of was that of Prof. Dryden at the Experiment Station of Utah. Prof. Dryden's work was of a demonstrative nature. His early bulletins were forceful and well illustrated, and did much to call attention to poultry work.

In all this early work the great Mississippi Valley, where four-fifths of the nation's poultry is produced, entirely ignored the hen. The writer began his work with poultry at the Kansas Station in 1902, but his chickens were housed in a discarded hog house, and no funds being available, little was accomplished. In the last three or four years these experiment stations are rapidly falling into line and a number of poultry bulletins have recently been issued from these younger schools.

A few of the early landmarks in experiment station work was as follows:

The Utah Station clearly found that hens laid about 65% as many eggs in the second as in the first year, and that to keep hens for egg production beyond the second year was unprofitable.

Massachusetts proved that corn was a better food for layers than wheat, and that the prejudice against it was founded on a misapplied theory.

The New York Station at Geneva demonstrated that poultry generally, and ducks in particular, are not vegetarians and must have meat to thrive, and that vegetable protein will not make good the deficiency.

The Maine Station was chiefly instrumental in introducing trap-nests, curtain-front houses and dry feeding. The breeding work at Maine will be discussed at length in the last section of this chapter.

The United States Department of Agriculture did not take up poultry work until 1906. The publications issued by the department before that time were written by outsiders and printed by the Government.

The following is the list of the addresses of the experiment stations who have taken a leading interest in poultry work. It is not worth while giving a list of poultry bulletins, as many of them are out of print and can only be consulted in a library.

Maine—Orono.	Iowa—Ames.
Mass.—Amherst.	Kansas—Manhattan.
Conn.—Storrs.	Utah—Logan.
Rhode Is.—Kingston.	Calif.—Berkeley.
New York—Ithaca.	Oregon—Corvallis.
New York—Geneva.	U. S. Gov.—Washington, D. C.
Maryland—College Park.	Ontario—Guelph (Canada).
West. Va.—Morgantown.	

Many foreign governments have us outdistanced in the encouragement of the poultry industry. Our Canadian neighbors have done much more practical work in getting out among the farmers and improving the stock and methods along commercial lines. As a result the Canadians have built up a nice British trade with which we have thus far not been able to compete. The work by the Ontario Station on the subject of incubation is discussed in the Chapter on Incubation.

Australia, like Canada, has given much practical assistance in marketing the poultry products, the government maintaining packing stations where the poultry is packed for export.

The Australian laying contests are quoted in the present volume. They outclass anything else in the world along that line.

In England, Ireland and especially in Denmark, the government, or societies encouraged by the Government, have done a great deal to develop the poultry industry. Depots for marketing and grading are maintained and the stock of the farmers is improved by fowls from the government breeding farms.

THE STORY OF THE "BIG COON"

With apologies to Joel Chandler Harris, I will tell a little story.

Uncle Remus was telling the little boy about the "big coon." It seems that the "big coon" had been seen on numerous occasions, but all efforts at his capture had failed. One night they saw the "big coon" up in the 'simmon tree, in the middle of the ten-acre lot. All hands and the dogs were summoned. To be sure of bagging the game, the tree was cut down. The dogs rushed in but there was no coon.

"But, Uncle Remus," said the little boy, "I thought you said you saw the big coon in the tree."

"Laws, chile," replied Uncle Remus, "doesn't youse know dat it am mighty easy for folks to see something dat ain't dar, when dey are lookin' fer it?"

When scientific experimenters entered the poultry field about fifteen years ago, they found it swarming with old ladies' notions. For everything a reason was given, but these reasons were derived from the kind of dreams where that which pleases the human mind is seized upon and search is made to find ideas to back it, not because it is true, but because it "listens good" to the dreamer. The first duty of the scientist was to banish these will-o'-the-wisp ideas that lead to no practical results.

For illustration: Round eggs were supposed to hatch pullets and long ones cockerels. Eggs will not hatch if it thunders. Shipped eggs must be allowed to rest before hatching, the drug store was the universal source of relief when the chickens became sick, and red pepper and patent foods were the egg foods par excellence. These things, thanks to the scientist, are no longer believed or regarded by well read poultrymen, and instead his attention has been turned to matters having a more happy relation to his bank account.[*]

In clearing away the useless popular notions, the scientists themselves have not been free from their influence, especially when they seemed to agree with accepted scientific theory. Many, indeed, are the 'coons in poultry science that have been seen because they were being looked for.

As a partial explanation it should be said that men available for scientific poultry work are very scarce. Poultrykeepers schooled in the University of the Poultry Yard have no conception of scientific methods, and would explain experimental results by a theory that would fail to fit elsewhere. The available scientists on the other hand are seldom poultrymen.

Among the first men to take up animal husbandry work of all kind, were the veterinarians. For years the only poultry publications put out by the U.S. Government were by veterinarians. These dust-covered volumes with their five color plates of the fifty-seven varieties of tapeworms, still rest on the shelves of public libraries, a monument to the time when the practical poultryman knew only things that weren't so, and the scientific poultryman knew only things that were useless.

The first general law that all experimenters should know—and the ignorance of which has caused and still causes the waste of the

---

* However, these folkloric ideas still "listen good" and are widely believed, while the scientific discoveries, being rather dull and having only truth to recommend them, are forgotten. I find that, when talking to people with small flocks, it's often as if the Twentieth Century had never happened at all.

major portion of experimental brains and money—we will call the "Law of Chance." Let the reader who is not familiar with such things take two pennies and toss them upon the table. They are both heads up. He tosses them again, one comes heads, the other tails. The third time repeats the second. The fourth both come tails. The law of chance says this is correct. Heads should appear 25%, tails 25%, and mixed 50% of the time. Now let the reader try this in a lot of twelve tosses. Does it prove the law? Try it again. Are all lots alike? Now pitch a hundred times, then pitch pennies all day. By night the law will be so near proven that the experimenter will be willing to concede its validity.

Now suppose the lots of twelve tosses each were lots of twelve hens, one Plymouth Rocks, the other Wyandottes—or one fed corn and the other wheat. The law of chance clearly proves that the larger the number of units, the nearer the theoretical truths will be the experimental results. Note, however, that small lots may by chance be as near the truth as large lots.

In practice two grave errors are made: First, conclusions are drawn from small lots compared with each other; second, conclusions are drawn from large lots compared with small lots. In the first case both may be off; in the latter case the small one may be off. Examples of the first error are to be found in the scores of contradicting breed and feed tests that were published in the early days of poultry research. The second error is exemplified in the Ontario experiments in incubation, to which reference has already been made.

Here is a further example of this error. From the fifth egg laying competition at the Hawkesberry Agricultural College in Australia, I copy the following (figure 15, page 221).

The ranking of Cuckoo Leghorns as first is a chance happening due to the small number; likewise the Black Leghorns had a streak of bad luck and received lowest place. To one not familiar with such work, the real significance of the table is that the Single Comb White Leghorns did the best work. A totaling of all other varieties gives 84 fowls with an average egg production of 170.5, which bears out the conclusion. As these birds were all kept in pens of six, we would expect to find the highest single pen to be

No. of Hens	Variety	Ave. Egg Yield.
6	Cuckoo Leghorn	190.16
30	Single Comb Brown Leghorn	177.00
138	Single Comb. White Leghorn	174.93
12	Rose Comb Brown Leghorn	173.50
12	Rose Comb White Leghorn	172.66
18	Buff Leghorn	160.55
6	Black Leghorn	138.33

Figure 15. Performance by breed.

White Leghorns, because, when compared with all other Leghorns taken together, they have both higher average production and a larger number of hens. This accords with the fact that as the highest single pen is found to be White Leghorns with an egg yield of 239 eggs.

The above illustrates another important phase of the laws of chance, which says that not only is the average likely to be nearer the theoretical average sought when the number is increased, but that the individual extremes will be more removed.

Important Experimental Results at the Illinois Station

From an Illinois Experiment Station report, the following is quoted:

"The stock used was Barred Plymouth Rock pullets. These pullets were a very uniform Barred Rock stock that had been bred as an individual strain for many years. They were practically the same age, and except for the factors mentioned were treated as uniformly as possible.

First Year's Results.		
No. Hens	Diet	Avg. Egg Yield.
10	Nitrogenous Diet	132.9
10	Carbonaceous Diet	128.4
10	Wet Mash	155.8
10	Dry Mash	111.4

"The results of the first test are somewhat surprising for it is generally believed that the nitrogenous diet is best for laying hens. The difference indicated in the first year's results was so light that it was decided to repeat the experiment the second year.

"As the wet mash is clearly proven to be superior, these hens were used the second year to compare meat meal with fresh cut bone.

Second Year's Result.		
No. Hens	Diet	Avg. Egg Yield.
10	Nitrogenous	142.2
10	Carbonaceous	134.5
10	Meat Meal	102.2
10	Green Cut Bone	128.9

"The results of the second year clearly indicate the great superiority of green cut bone as compared with the dry unpalatable meat meal. The comparison of a highly nitrogenous ration with that of a ration consisting largely of corn, while showing the advantages of the nitrogenous rations, does not show the contrast expected.

"Some visiting poultrymen expressed the opinion that corn is a better poultry food than commonly supposed. Considering this fact and the great fundamental importance of the question at issue, it was decided to repeat the experiment a third year, and feed a large number of birds on each ration:"

No. Hens	Diet	Avg. Egg Yield.
100	Nitrogenous	126.9
100	Carbonaceous	127.2

I will leave the last without comment, for the whole thing is a hoax. The Illinois Experiment Station has never owned a chicken. These "Illinois" experiments were planned and executed in a few minutes of the writer's spare time. The basis of the experiments was a pack of cards containing the individual records of the Maine Experiment Station hens. Shuffling the cards and averaging the desired number of records as they come in the pack made the distinction between the various diets.

EXPERIMENTAL BIAS

Pet ideas consciously or unconsciously mold practice. A bias toward an idea may show itself in the planning and conducting of an experiment, or it may come out in the later interpretation.

An illustration of the first kind is found in the early work of the West Virginia Station (Bulletin 60). With the preconceived notion that hens should have a nitrogenous diet an experiment was planned and conducted as follows:

One lot of hens was fed corn, potatoes, oats and corn meal. A contrasted lot reveled in corn, potatoes, hominy feed, oat meal, corn meal and fresh cut bone. The results were in favor of the latter ration by a doubled egg yield.

To any experienced poultryman the reason is evident. The variety of the diet and the meat food are what made the showing.

About the same time the Massachusetts Station planned a similar experiment. The bias was the same, but it took a fairer form. The hens were both given a decent variety of food and some form of meat. The bulk of the grain was corn in the carbonaceous, and wheat in the nitrogenous ration. The results were in favor of the corn. This astonished the experimenter. He tried it again and again tests came out in favor of corn. At last the old theory was revoked, and the fallacy of wheat being essential to egg production was exploded. If by an irony of fate in the shuffling of the hens, the wheat pen had the first time showed an advantage, the experimenter might have been satisfied, and the waste of feeding high-priced feed, when a better and a cheaper is at hand, might have gone on indefinitely.

Of bias in the interpretation of results all publications are more or less saturated. A reading of the Chapter on Incubation will illustrate this. A common error of this kind is the omission of facts necessary to fully explain results. Items of costs are invariably omitted or minimized. Food cost alone is usually mentioned in figuring experiment station poultry profits, which statement will undoubtedly cause a sad smile to creep over the face of many a "has-been" poultryman.

The writer remembers an incident from his college days which illustrates the point in hand. Let it first be remarked that this was on the new lands of the trans-Missouri Country, where manure had no more commercial value than soil, and is freely given to those who will haul it away.

The professor at the blackboard had been figuring up handsome profits on a type of dairying towards which he was very partial. The figures showed a goodly profit, but the biggest expense item—that of labor—was omitted. One of the students held up his hand and inquired after the labor bill.

"Oh," said the smiling professor, "The manure will pay for the labor."

When the class adjourned, the student remarked: "They say figures won't lie, but a liar will figure."

The third way in which experiments are made worthless is by the introduction of factors other than the one being tested. This may be done by chance, and the experimenter not realize the presence of the other factor, or the varying factors may be introduced intentionally under the belief that they are negligible. Of the first

224

case an instance may be cited of the placing of two flocks in a house, one end of which is damper than the other, the accidental introduction into one flock of a contagious disease, or one flock being thrown off feed by an excessive feed of greens, etc., etc. These factors that influence pens of birds greatly add to the error of the law of chance. In fact it amounts to the same thing on a larger scale. For this reason not only are many individuals, but many flocks, many locations, and many years needed to prove the superiority of the contrasted methods.

The criticisms in the following section will amply illustrate the case of foreign factors being unwisely introduced into an experiment.

THE EGG BREEDING WORK AT THE MAINE STATION

As is well known, the Maine Station was for years considered by all poultrymen to be doing a great and beneficial work in breeding for increased egg production. Up until the fall of 1907, the poultrymen of the country were of the opinion that this work was in every way successful, and a large number of private breeders had taken up the use of trap-nests in an effort to build up the egg production of their fowls.

When early in 1908 Bulletin 157 of the Maine Experiment Station was published, it showed by averages, as given in the table on page 226, that the egg yield at the station was for the entire period on the decline. In Bulletin 157, the statement was made that "arithmetical mistakes" and "faulty statistical methods" accounted for the discrepancies between the former publications and the criticized data. The further explanation that "the experiment was a success as an experiment," etc., only appeared to the public mind as a graceful way of explaining what was, to the practical man, an utter failure of the entire work.

The unfortunate death of Professor Gowell, together with the fact that he had equipped a private poultry farm with station stock, added to the confusion, and the result of the bulletin was the precipitation of a general "pow-wow" in which the poultry

Year	Hens in Flock	Floor Area per Hen	Egg Yield per Hen
1900	20	8. sq. ft.	136.36
1901	20	8. sq. ft.	143.44
1902	20	8. sq. ft.	155.58
1903	20	8. sq. ft.	135.42
1904	50	4.4 sq. ft.	117.90
1905	50	4.4 sq. ft.	134.07
1906	50	4.4 sq. ft.	140.14
1907	50	4.4 sq. ft.	113.24

Figure 16. Results of the Maine Experiment Station

editors were about equally divided between those who were casting insinuations upon the personnel of the station and those who decried the whole effort toward improving the egg yield.

After going over the publications of Professor Gowell, visiting the station and meeting the present force, I came to the following conclusions regarding the matter:

Professor Gowell's work is open to severe criticism. Errors have been made in conducting the work at Maine which have made it possible for a mathematical biologist to take the data and seemingly prove that selection, as practiced by Professor Gowell, actually resulted in lowering inherent egg capacity of the strain of Plymouth Rock hens under experimentation. Had Professor Gowell's successor been a practical poultryman, it is my candid opinion that the public would have been given a radically different explanation of the results.

Professor Gowell is the author of the following statement: "The small chicken grower is earnestly urged to use an incubator for hatching." This opinion is not in accord with that of the majority of breeders and the more progressive experiment station workers. The opinion has been expressed by Professor Graham and others, that the particular results at the Maine Station may have been due to the decrease of vitality caused by continued artificial hatching. This view may be wholly without foundation.

Nevertheless, as the common type of incubator is under heavy criticism, and it is pretty well proven that chicks so hatched have not the vitality of naturally hatched chicks, surely a series of breeding experiments would carry more weight if the replenishing of the flock had been accomplished by natural means.

For the first few years of the breeding work the house used was the old-fashioned double-walled and warmed pattern. The last few years of this work were conducted in curtain-front houses. That the cool house is an improvement over the warm house is generally conceded, but there are many poultrymen who are still of the opinion that the warm house will give a larger egg yield, though at a greater expense and less profit.

In the early years of the work the method of feeding was also a time-honored one, and included a warm wet mash. About the middle of the experimental period Professor Gowell brought out the system of feeding dry mash from hoppers. This custom became a great fad and Professor Gowell and Director Woods have preached it far and wide. Perhaps it is an improvement, but it is today much more popular with novices than with established egg farms. Many old line poultrymen have tried dry mash only to go back to wet mash, by which method the hens can be induced to eat more, which is conducive to high egg yields. Whether these changes in housing and feeding have been improvements as claimed by those who introduced them, or whether their popularity may be explained in part at least by the psychology of fads, is a point in question, but certainly the marring of a breeding experiment by introducing radical changes in the factors of production is, at best, unfortunate.

A much more serious criticism than any of the foregoing is to be found in a change of the size of flocks and amount of floor space per fowl. I have gone over carefully the published records of Professor Gowell, and the review of Dr. Pearl, and figure 16 represents, as near as I can determine, these factors for the series of years. In the year 1903 I find no clear statement as to the manner in which the birds were housed, and I may be in error in this case. Otherwise the table gives the facts.

Certainly this oversight is a serious one, and one especially remarkable considering the fact that the comparison of different size flocks formed a prominent part of the Maine Station work during the last three years of the breeding test. The results of the work at the Maine Station on testing flock size, conducted without relation to the breeding work, gave the following results:

No. of Hens	Sq. ft. per Hen	Egg Yield
150	3.2	111.68
100	4.8	123.21
50	4.8	129.69

Figure 17. Effect of flock size and crowding on egg yield.

No comparisons of 50 and 20 bird flocks in the same year are available, but by extending the comparisons of the 50, 100 and 150 flocks into the 20 flock size, we can get some idea of the error that has been here introduced. The result of the Australian egg laying contest, in which the flocks were composed of six hens, shows a yield of about one and one-half times as heavy as the Maine records, which certainly seems to substantiate the ideas here brought out.

It is a well-established fact in poultry circles that many men who succeed with a few hundred hens fail when the number is increased to as many thousands. When the breeding experiments under discussion were started, Professor Gowell had under his supervision about three hundred hens. When the work was closed the experiment station plant had been increased to four or five times its capacity, and Professor Gowell had a large private poultry plant of his own in addition.

It is interesting to note in this connection that the last four years of the records are explained by Professor Gowell as being low, due to various "accidents"(?) It is unreasonable to suppose the true explanation of these "accidents" would be found in connection with the increased responsibility and size of the plant.

The breeding stock sent out by Professor Gowell has given general satisfaction, and was found by Professor Graham of the Ontario Station, as well as by a number of private individuals, to be of superior laying quality to that of the average Barred Rock.

Clearly there is only one way to prove whether Professor Gowell's work has been a wasted effort, and that is for flocks of his strain to be tested at other experiment stations against birds of miscellaneous origin.

That much has been lost to the poultrymen of the country by the recent upheaval at the Maine Station, I believe to be the case, but that does not mean that the men now in charge will not in the future be of great value to the poultry interests. They are, however, in the class of pure scientists rather than applied scientists, but if let alone they will dig out something sooner or later which they or others can apply to the benefit of the industry.

Upon the whole, I think that the present case of the trap-nest method of increasing egg production stands very much as it is has always stood, being a commendable thing for small breeders who could afford the time, but not practical in a large way, except at experiment stations. On a large commercial scale the system of selecting sires by the collective work of his first year's offspring would probably get the quickest results.[*]

The best use of the funds of the people in the promotion of agricultural industries is in the permanent endorsement on the one hand of a few high-grade research stations where the deeper theories may be worked out, and on the other the teaching of such good principles and practices as are already known.

The greatest opportunity for Government effort lies in the development demonstration farm work in poultry, just as it is doing with corn and cotton in the South.

---

* This is the method that eventually won out.

Chapter 16. Poultry On The General Farm

This chapter will be devoted to specific directions for the profitable keeping of chickens on the typical American farm. By "typical American farm" I mean the farm west of Ohio, north of Tennessee and east of Colorado. Farms outside this section present different problems. In the region mentioned about three-quarters of the American poultry and egg crop is produced, and in this section poultrykeeping is more profitable when conducted as a part of general farm operations than as an exclusive business.

There is no reason why a farmer should not be a poultry fancier if he desires, but in that case his special interest in his chickens would throw him out of the class we are at present considering. Likewise, I do not doubt that in many instances where the farmer or members of his family took special interest in poultry work, it would be profitable to increase the size of operations beyond those herein advised, using incubators and keeping Leghorns. Of these exceptions the farmer himself must judge. The rules I lay down are for those farmers who wish to keep chickens for profit, but do not care to devote any larger share of their time and study to them than they do to the cows, hogs, orchard or garden.

The advice herein given in this chapter will differ from much of the advice given to farmers by poultry writers. The average poultry editor is afraid to give specific advice concerning breeds, incubators, etc., because he fears to offend his advertisers. The reader, left to judge for himself, is liable to pick out some fancy impractical variety or method.

BEST BREEDS FOR THE FARM

Keep only one variety of chickens. Do not bother with other varieties of poultry unless it is turkeys. Whether it will pay to raise turkeys will depend upon your success with the little turks, and on the freedom of the community from the disease called blackhead.

The kind of chicken you should keep should be picked from the three following breeds: Barred Plymouth Rock, White Wyandotte, or Rhode Island Reds. If you go outside of these three breeds be sure you have a very good reason for doing so.

To get a start with a new breed, buy at least four sittings of eggs in a single season, paying not over $2.00 per sitting. Keep all the pullets and a half dozen of the best cockerels. The next spring pen these pullets up with the best cockerels, and use none but eggs from this pen for hatching. That fall, sell all of the young cockerels and all the old scrub hens. The second spring the two old roosters from the original purchased eggs are used with the general flock. From this time on the entire flock is pure-bred and should remain so.

Each year, when the chicks are about six or eight weeks old, pick out the largest, most vigorous male chick from each brood. Mark these by clipping the web of the foot or putting on leg bands. From those so marked the breeding cockerels for the next season are later selected. When you pick the good cockerels, pick out all runty-looking pullets and cut off the last joint of the hind toe. These runts are later to be eaten or sold. The more surplus chicks raised, the more strictly can the selection be made.

This system of picking the best cockerel from each brood and discarding the poorest pullets is the most practical method known of building up a vigorous, quick-growing and early laying strain.

When we allow the entire flock of many different ages to grow up before the selection is made, it is impossible to select intelligently.

Every third or fourth year an extra cock bird may be purchased provided you are sure you are getting a specimen from a better flock than your own. Swapping roosters or eggs every year is poor policy. If your neighbor has better stock than you, get his blood pure and sell off your own, but do not keep a barnyard full of scrubs who can trace their ancestry to every flock in the neighborhood.

KEEP ONLY WORKERS

On many farms, few eggs are gathered from October to January. This is a season when eggs bring the best prices. To secure eggs at this season, the first requisite is that the pullets be hatched between the first of March and the middle of May, or, in the case of Leghorns, between the first of April and the first of June. Pullets hatched later than these dates are a source of expense during the fall and early winter. On the other hand, it is an unnecessary waste of effort to hatch pullets before the dates mentioned, because, if hatched too early, they will molt in the fall and stop laying the same as old hens.[*]

Pullets must be well fed and cared for if expected to develop in the time allowed. As they begin to show signs of maturity they should be gotten into permanent quarters. If allowed to begin laying while roosting in coops or in trees they will be liable to quit when changed to new quarters. If possible the coops should be gradually moved toward the hen-house and the pullets gotten into quarters without excitement or confinement. The poultry-house should have an ample circulation of fresh air. Young stock that have been roosting in open coops are liable to catch cold if confined in tight houses.

A common mistake is to allow a large troop of young roosters to overrun the premises in the early fall. Not only is money lost in the decrease in price that can be obtained for these cockerels, but the pullets are greatly annoyed, to the detriment of the egg yield.

Any chicken that is not paying for its food in growth or in egg production is a source of loss. As soon as the hatching season is over, old roosters should be sent to market. In June through August, egg production is not very profitable, and a thorough culling of the hens should be made. Market all hens two years or more

---

[*] I have learned, to my cost, that this is still as true as ever.

of age. Send with these all the yearling hens that appear fat and lazy. By the time the young pullets are ready to be moved into quarters—the latter part of August—these hens should be reduced to about one-half the original number. Sometime during September a final culling of the old stock should be made. Those that have not yet begun to molt should be sold, as they will not be laying again before the warm days of the following February. This system of culling will leave the best portion of the yearling hens, which, together with the early-hatched pullets, will make a profitable flock of layers.

HATCHING CHICKS WITH HENS

The eggs for hatching should be stored in a cool, dry location at a temperature between forty and seventy degrees Fahrenheit. A good rule is not to set eggs over two weeks old.

The two chief losses with sitting-hens are due to lice and interference of other hens. The practice of setting hens in the chicken-house makes both these difficulties more troublesome. Almost all farms will have some outbuilding situated apart from the regular chicken-house that can be used for sitting hens. The most convenient arrangement will be to use boxes, and have these open at the top. They may be placed in rows and a plank somewhat narrower than the boxes used as a cover. The nests should be made by throwing a shovel of earth into the box and then shaping a nest of clean straw. Make the nest roomy enough so that as the hen steps into the nest the eggs will spread apart readily and not be broken. When a hen shows signs of broodiness remove her to the sitting-room. This should be done in the evening, so that the hen becomes accustomed to her position by daylight. Place the hen upon the nest-eggs and confine her to the nest. If all is well the next evening give her a full setting of eggs.

A practical method to arrange for sitting-hens is to build the nests out of doors, allowing each hen a little yard, so that she may have liberty to leave her nest as she chooses. These nests may be built by using twelve-inch boards set on edge, so as to form a series of small runways about one by six feet. In one end are built the nests, which are covered by a broad board, while the remainder of the arrangement is covered with lath or netting. The food, grit and water should be placed at the opposite end of the runway. Care should be taken to locate these nests on well-drained ground. Arrangements should be made to close the front of the nest during hatching so that the chicks will not drop out. A contrivance of this kind furnishes a very convenient method of handling sitting-hens, and if no separate building is available would be the best method to use.

INCUBATORS ON THE FARM

My candid advice to the farmer who is in doubt as whether to buy an incubator or not, is to let it alone. If the farmer reads the chapter on artificial incubation, he will see that he is dealing with a very complex problem, and one in which his chances of success are not very great.

In order to learn the facts concerning incubators on the farms, the writer made a special investigation on the subject while poultryman at the Kansas Experiment Station. Replies received from 111 Kansas farmers, report 21 as having tried incubators. Of these, 6 reported the incubators as being an improvement over hatching with hens; 10 reported the incubator as being successful, but not better than hens, while the remaining 5 declared the incubator to be a failure. The results of this inquiry, and of personal visits to farms, led the writer to believe that about one-tenth of the farmers of Kansas had tried incubators, and that about as many failed as succeeded with artificial hatching.

The argument for the incubator on the farm is certainly not one of better hatching, but there is an argument, and a good one for the farm incubator. The argument is this:

Hens will not set early enough and in sufficient quantities to get out as large a number of chicks as the farmer may desire. Now, each hen will not hatch over 10 chicks, but is capable of caring for at least 30. Here the incubator comes into good use, for the farmer can set a half dozen hens along with the incubator, and give all the chicks to the hens. This is the method I recommend where an incubator is to be used. The development of the public hatchery would supply these other 20 chicks more economically and more certainly than the farm incubator, but until that institution becomes established the more ambitious farm poultry raisers are justified in trying an incubator.

The best known incubators in the market are the Cyphers, the Model and the Prairie State. Cheaper machines are liable to do poor work.

The following points may help the farmer in deciding whether or not to buy an incubator and in picking out a good machine:

The person to run the incubator is the first condition of its success. A good incubator requires attention twice a day. One person should give this attention, and must give it regularly and carefully. The farmer's wife or some younger member of the family can often give more time and interest to this work than can the farmer. The likelihood of a person's success with artificial hatchers can best be determined by himself.

The best location for an incubator is a moderately damp cellar. The next choice would be a room in the house away from the fire or from windows. Drafts of air blowing on the machine are especially to be avoided. Not only do they affect the temperature directly, but cause the lamp to burn irregularly, and this may result in fire.

The objects in view in building an incubator are: (1) To keep the eggs at a proper temperature (103° on a level with the top of the eggs). (2) To cause the evaporation of moisture from the eggs at a normal rate. (3) To prevent the eggs from resting too long in one position.

The case of the incubator should be built double- or triple-walled to withstand variation in the outside temperature. The doors should fit neatly and be made of double glass. The lamp should be made of the best material, and the wick of sufficient width that the temperature may be maintained with a low blaze. The most satisfactory place for the lamp is at the end of the machine, outside the case.

Regulators composed of two metals, such as aluminum and steel, are best. Wafers filled with ether or similar liquid are more sensitive but weaker in action. Hard rubber bars are frequently used.

The most practical system of controlling evaporation is a system of forced ventilation, in which the air is heated around the lamp-flue and passes through the egg-chamber at a rate determined by ventilators in the bottom of the machine. With the outside air cold and dry only slight current is required, but as the outer air becomes warmer or damper more circulation is needed.

Turning the egg is not the work that many imagine it to be. It is not necessary that the egg be turned with absolute precision and regularity. An elaborate device for this work is useless. The trays will need frequently to be removed and turned around or shifted, and the eggs can be turned at this time by lifting out a few on one side of the tray and rolling the others over.

Two other points to be considered in the incubator are a suitable nursery or place for the newly hatched chick, and a good thermometer.

REARING CHICKS

If it is very early in the spring and the ground is damp, it is best to put the hen and her brood in some building. During the most of the season the best thing is an outdoor coop. The first consideration in making a chicken-coop is to see that it is rain-proof and rat-tight. The next thing to look for is that the coop is not air-tight. Let the front be of rat-tight netting or heavy screen. The same general plan may be used for small coops for hens, or for larger coops to he used as colony-houses for growing chickens.

The essentials are: A movable floor raised on cleats, a sliding front covered with rat-tight netting, and a hood over the front to keep the rain from beating in. If used late in the fall or early in the spring a piece of cloth should be tacked on the sliding front.

The chicken-coops should not be bunched up, but scattered out over as much ground as is convenient. Neither should they remain long in one spot, but should be shifted a few feet each day. At first water should be provided at each coop, but as the chicks grow older they may be required to come to a few central water pans.

As before suggested, rearing chicks with hens is the only suitable method for general farm practice. The brooder on the farm is an expensive nuisance.

For brooder-raised chicks it is necessary to provide means for the little chick to exercise. But in the season when the vast majority of farm chicks are raised they may be placed out of doors from the start and the trouble will now be to keep them from getting too much exercise; i.e., to keep the hens from chasing around with them, especially in the wet grass. This is properly prevented by keeping the brood coops in plowed ground, and keeping the hens confined by a slatted door until the chicks are strong enough to follow her readily.

The chick should not be fed until 48 to 72 hours old.[*] It may then be started on the same kind of food as is to form its diet in later life. The hard boiled egg and bread and milk diets are wholly unnecessary and are only a waste of time.

I recommend the same system of chick feeding for the general farm as is used on commercial plants, and I especially insist that it will pay the farmer to provide meat food of some sort for his growing chicks. The amount eaten will not be large, nor need the farmer fear that supplying the chicks with meat food will prevent their consuming all the bugs and worms that come their way.

Besides comfortable quarters, the chick to thrive, must have exercise, water, grit, a variety of grain food, green or succulent food, and meat food.

Water should be provided in shallow dishes. This can best be arranged by having a dish with an inverted can or bottle which allows only a little water to stand in the drinking basin.

Chicks running at large on gravelly ground need no provision for grit. Chicks on board floors or clay soils must be provided with either coarse sand or chick grit, such as is sold for the purpose.

---

[*] Current practice is to feed chicks within a few hours of hatching.

Grain is the principal and too often the only food of the chick. The common farm way of feeding grain to young chickens is to mix corn-meal and water and feed in a trough or on the ground. There is no particular advantage in this way of feeding, and there are several disadvantages. The feed is all in a bunch, and the weaker chicks are crowded out, while if wet feed is thrown on the ground or in a dirty trough the chicks must swallow the adhering filth, and if any food is left over it quickly sours and becomes a menace to health. Some people mix dough with sour milk and soda and bake this into a bread. The better way is to feed all of the grain in a natural dry condition.

There are foods in the market known as chick foods. The commercial foods contain various grains and seeds, together with meat and grit. Their use renders chick feeding quite a simple matter, it being necessary to supply in addition only water and green foods. For those who wish to prepare their own chick foods the following suggestions are given:

Oatmeal is probably the best grain food for chicks. Oats cannot be suitably prepared, however, in a common feed-mill. The hulled oats are what is wanted. They can be purchased as the common rolled oats, or sometimes as cut or pin-head oatmeal. The latter form would be preferred, but either of these is an excellent chick feed. Oats in these forms are expensive and should be purchased in bulk, not in packages. If too expensive, oats should be used only for a few days, when they may be replaced by cheaper grains. Cracked corn is the beet and cheapest chick food. Flaxseed could be used in small quantities. Sorghum, wheat, cow-peas—in fact any wholesome grain—may be used, the more variety the better. Farmers possessing feed-mills have no excuse for feeding chicks exclusively on one kind of grain. If there is no way of grinding corn on the farm, oatmeal, millet seed and corn chop can be purchased. At about one week of age whole sorghum, and, a little later whole wheat, can be used to replace the more expensive feeds.

Green or bulky food of some kind is necessary to the healthy growth of young chickens. Chickens fed in litter from clover or alfalfa will pick up many bits of leaves. This answers the purpose fairly well, but it is advisable to feed some leafy vegetable, as kale or lettuce. The chicks should be gotten on some growing green crop as soon as possible.

Chickens are not by nature vegetarians. They require some meat to thrive. It has been proven in several experiments that young chickens with an allowance of meat foods make much better growth than chickens with a vegetable diet, even when the chemical constituents and the variety of the two rations are practically the same.

Very few farmers feed any meat whatever. They rely on insects to supply the deficiency. This would be all right if the insects were plentiful and lasted throughout the year, but as conditions are it will pay the farmer to supplement this source of food with the commercial meat foods.

Fresh bone, cut by bone-cutters, is an excellent source of the meat and mineral matter needed by growing chicks. If one is handy to a butcher shop that will agree to furnish fresh bones at little or no cost, it will pay to get a bone-mill, but the cost of the mill and labor of grinding are considerable items, and unless the supply of bones is reliable and convenient this source of meat foods is not to be depended upon.

The best way to feed beef-scrap is to keep a supply in the hopper so the chickens may help themselves. In case meat food is given, bone-meal, fed in small quantities, will form a valuable addition to their ration. Infertile eggs from incubators, as well as by-products of the dairy, can be used to help out in the animal-food portion of the ration. Chickens may be given all the milk they will drink. It is generally recommended that this be given clabbered.

FEEDING LAYING HENS

The food requirements of a laying hen are very like those of a growing chicken. One addition to the list is, however, required for egg production, which is lime, of which the shell of the egg is formed. In the summer-time hens on the range will find sufficient lime to supply their needs. In the wintertime they should be supplied with more lime than the food contains. Crushed oyster shell answers the purpose admirably.[*]

A supply of green food is one of the requisites of successful winter feeding. Every farmer should see that a patch of rye, crimson clover, or some other winter green crop is grown near his chicken-house. Vegetables and refuse from the kitchen help out in this matter, but seldom furnish a sufficient supply. Vegetables may be grown for this purpose. Mangels and sugar-beets are excellent. Cabbage, potatoes and turnips answer the purpose fairly well. Mangels are fed by splitting in halves and sticking to nails driven in the wall.

Clover and alfalfa are excellent chicken feeds and should be used in regions where winter crops will not keep green. The leaves that shatter off in the mow are the choicest portion for chicken feeding, and may be fed by scalding with hot water and mixing in a mash. Hens will eat good green alfalfa if fed dry in a box.

The feeding of sprouted oats should be practiced when no other green food is available. Oats may he prepared for this purpose by thoroughly soaking in warm water and being kept in a warm, damp place for a few days. Feed when the sprouts are a couple of inches long.

Almost all grains are suitable foods for hens. Corn, on account of its cheapness and general distribution, is the best. The general prejudice against corn feeding should be directed rather against feeding one grain alone without the other forms of food. If hens are supplied with green feed, with mineral matter, some form of meat food, and are forced to take sufficient amount of exercise, then overfatness due to the feeding of corn need not be feared.

As has already been emphasized, the variety of food given is more essential than the kind. Do not feed one grain all the time. The more variety fed the better. Corn and sorghum, being cheap grains, will form the major portion of the ration, but even if much

------

[*] Hens really ought to be given oystershell year-round.

higher in price, it will pay to add a portion of such grains as wheat, barley, oats or buckwheat.

CLEANLINESS

The advice commonly given in poultry papers would require one to take nearly as many pains in the cleaning of a chicken house as in the cleaning of a kitchen. Such advice may be suitable for the city poultry fanciers, but it is out of place when given to the farmer. Poultry raising, the same as other farm work, must pay for the labor put into it, and this will not be the case if attempt is made to follow all the suggestions of the theoretical poultry writer. *

The ease with which the premises may be kept reasonably free from litter and filth is largely a matter of convenient arrangement. The handiest plan from this view-point is the colony system. In this the houses are moved to new locations when the ground becomes soiled. If the chicken-house is a stationary structure it should be built away from other buildings, scrap-piles, fence corners, etc., so that the ground can be frequently freshened by plowing and sowing in oats, rye or rape. The ground should he well sloped, so that the water draining from the surface may wash away much of the filth that on level ground would accumulate.

Cleanliness indoors can be simplified by proper arrangement. First, the house must be dry. Poultry droppings, when dry, are not a source of danger if kept out of the feed. They should be removed often enough to prevent foul odors. Drinking vessels should be rinsed out when refilled and not allowed to accumulate a coat of slime. If a mash is fed, feed-boards should be scraped off and dried in the sun. Sunshine is a cheap and efficient disinfectant.

---

* I've found this advice to be very true. Having respect for the value of your own labor is the biggest difference between the practical farmer and the hobbyist, who will often complicate matters as an excuse to spend more time on his hobby.

The advice on the control of lice and the method of handling sick chickens that has been given in the main section of the book, will apply as well on the farm as on the commercial poultry plant. Certainly the farmer's time is too valuable to fool with the details of poultry therapeutics.

FARM CHICKEN HOUSES

The following notes on poultry houses apply to Iowa and Nebraska, where the winters are severe, and similar climates. Farther south and east the farmer should use the same style of houses as recommended for egg farms. A chicken house just high enough for a man to walk erect and a floor space of about 3 square feet per hen is advisable. This requires a house 12 by 24 for one hundred hens, or 10 by 16 for fifty (roughly three square feet per hen).

Lands sloping to south or southeast, that dries quickly after a rain, will prove the most suitable for chickens. A gumbo patch should not be selected as a location for poultry. Hogs and hens should not occupy the same quarters, in fact, should be some distance apart, especially if heavy breeds of chickens are kept.[*] Hens should be removed from the garden, but may be near an orchard. Chicken-houses should be separated from tool-houses, stables, and other outbuildings.[†]

Grading the ground for chicken-houses is not commonly practiced, but this is the easiest means of preventing dampness in the house, and is necessary in heavy soils. The ground level may

---

[*] Not only will pigs turn chicken areas into a wallow, but they quickly learn to kill and eat chickens.

[†] At one time we had chickens roosting in our barn and car port, causing no end of trouble. Among other things, they carried a serious mite infestation up with them into their roosting sites in the rafters, where it was difficult to deal with. The mites spread from there into the henhouses and brooder houses. It was a happy day when we caught all the rafter-roosting hens and took them off to the auction yard.

be raised with a plow and scraper, or the foundation of the house may be built and filled with dirt.

A stone foundation is best, but where stone is expensive may be replaced by cedar, hemlock or Osage orange posts, deeply set in the ground. Small houses can be built on runners as described for colony houses for an egg farm.

Floors are commonly constructed of earth, boards or cement. Cement floors are perfectly sanitary and easy to keep clean. The objections to their common use is the first cost of good cement floors. Cheaply constructed floors will not last. Board floors are very common and are preferred by many poultrymen, but if close to the ground they harbor lice, while if open underneath they make the house cold. Covering wet ground by a board floor does not remedy the fault of dampness nearly so effectually as would a similar expenditure spent in raising the floor and surrounding ground by grading. All things considered, the dirt floor is the most suitable. This should be made by filling in above the outside ground-level. The drainage will be facilitated if the first layer of this floor be of cinders, small rocks or other coarse material. Above this layer should be placed a layer of clay, wet and packed hard, so the hens cannot scratch it up, or a different plan may be used and the floor constructed of a sandy or loamy soil of which the top layer can be renewed each year.

The walls of a chicken-house must first of all be wind-tight. This may be attained in several ways. Upright boards with cracks battened is the cheapest method. Various kinds of lap-siding give similar results. The single-board wall may be greatly improved by lining with building-paper. This should be put on between the studding and siding. Lath should also be used to prevent the paper bagging out from the wall. The double-board wall is the best where a warm house is desired. It should be made by siding up outside the studding with cheap lumber. On this is placed a layer of roofing paper and over it the ordinary siding.[*]

---

[*] Plywood or corrugated metal sheathing gives a wind-tight wall in a single layer, especially if you caulk the seams.

The windows of a chicken-house should furnish sufficient light that the hens may find grain in the litter on cloudy days. Too much glass in a poultry house makes the house cold at night, and it is a needless expenditure.

The subject of roofing farm buildings may be summarized in this advice: Use patent roofing if you know of a variety that will last; if not, use shingles.[*] Shingle roofs require a steeper pitch than do roofs of prepared roofing. A shingle roof can be made much warmer by using tightly laid sheathing covered with building-paper. Especial care should be taken that the joints at the eaves of the house are tightly fitted.

The object of ventilating a chicken-house is to supply a reasonable amount of fresh air, and, equally important, to keep the house dry. Ventilation should not be by cracks or open cupolas. Direct drafts of air are injurious, and ventilation by such means is always the greatest when least needed.

Schemes of ventilation by a system of pipes are expensive and unnecessary. The latest, best and cheapest plan for providing ventilation is the curtain front house for the north, and the open front house for the more southerly sections. The curtain front house is giving way to the open front with a somewhat smaller opening in sections as far north as Connecticut.[†]

Make all roosts on the same level. The ladder arrangement is a nuisance and offers no advantage. Arrange the roosts so that they may be readily removed for cleaning. Do not fill the chicken-house full of roosts. Put in only enough to accommodate the hens, and let these be on one side of the house. The floor under the

---

[*] This was written before the widespread use of galvanized steel roofing, which is the most common roof these days.

[†] The curtain-front house used an open-weave cloth, such as cheesecloth, to allow airflow while not allowing gusts to enter the house. As implied by Hastings, curtains were already falling out of favor in 1909. The experience of many poultrymen was that when the curtains rotted away after a couple of winters and the farmer did not replace them right away, nothing bad happened.

roosts should be separated from the feeding floor by a board set on edge.

For laying flocks the nests must be clean, secluded and plentiful. Boxes under the roost-platform will answer, but a better plan is to have the nests upon a shelf along a side wall so arranged as to allow the hen to enter from the rear side. Nests should be constructed so that all parts are accessible to a white-wash brush. The less contrivances in a chicken-house, the better.

The farmer can get along very well without any chicken yard at all. It will, however, prove a very convenient arrangement if a small yard is attached to the chicken house. The house should be arranged to open either into the yard or out to the range. This yard may be used for fattening chickens or confining cockerels, or perhaps to enclose the flock during the ripening of a favorite tomato or berry crop.

THE END

For More Information

Or to order copies of our books over the Internet, see the Norton Creek Press Web Page at:

http://www.plamondon.com/nortoncreekpress.html.

Also From Norton Creek Press

SUCCESS WITH BABY CHICKS
BY ROBERT PLAMONDON

In this engaging, practical book, Robert Plamondon describes how to raise baby chicks, using brooder-house techniques that, though now largely forgotten, were once used on millions of American farms. Anyone can raise baby chicks successfully by following these simple, proven methods.

155 Pages, $13.95 ISBN 0-9721770-0-0

Coming Soon From Norton Creek Press

FEEDING POULTRY
BY G. F. HEUSER

A classic work on poultry nutrition, written for both the practical farmer and the serious student. This work, which dates from the 1950s, was written in an era when poultry nutrition was up to date, but when poultry nutritionists were still familiar with free range and small-farm feeding concerns.

AVAILABLE IN APRIL, 2003